Waste Management and the Environment X

WITPRESS

WIT Press publishes leading books in Science and Technology.
Visit our website for the current list of titles.
www.witpress.com

WITeLibrary

Home of the Transactions of the Wessex Institute.
Papers contained in this volume are archived in the WIT eLibrary in volume 247 of WIT Transactions on Ecology and the Environment (ISSN 1743-3541).
The WIT eLibrary provides the international scientific community with immediate and permanent access to individual papers presented at WIT conferences.
Visit the WIT eLibrary at www.witpress.com.

TENTH INTERNATIONAL CONFERENCE ON
WASTE MANAGEMENT AND THE ENVIRONMENT

Waste Management 2020

Conference Chairmen

Juan Casares
University of Santiago de Compostela, Spain
Member of WIT Board of Directors

Hideaki Itoh
Nagoya University, Japan

Massimiliano Lega
University of Naples Parthenope, Italy

International Scientific Advisory Committee

Organised by
Wessex Institute, UK
Parthenope University of Naples, Italy
Nagoya University, Japan

Sponsored by
WIT Transactions on Ecology and the Environment
International Journal of Environmental Impacts

WIT Transactions

Wessex Institute
Ashurst Lodge, Ashurst
Southampton SO40 7AA, UK

Senior Editors

Editorial Board

Waste Management and the Environment X

Editors

Juan Casares
University of Santiago de Compostela, Spain
Member of WIT Board of Directors

Hideaki Itoh
Nagoya University, Japan

Massimiliano Lega
University of Naples Parthenope, Italy

WIT*PRESS* Southampton, Boston

Editors:

Juan Casares
University of Santiago de Compostela, Spain
Member of WIT Board of Directors

Hideaki Itoh
Nagoya University, Japan

Massimiliano Lega
University of Naples Parthenope, Italy

Published by

WIT Press
Ashurst Lodge, Ashurst, Southampton, SO40 7AA, UK
Tel: 44 (0) 238 029 3223; Fax: 44 (0) 238 029 2853
E-Mail: witpress@witpress.com
http://www.witpress.com

For USA, Canada and Mexico

Computational Mechanics International Inc
25 Bridge Street, Billerica, MA 01821, USA
Tel: 978 667 5841; Fax: 978 667 7582
E-Mail: infousa@witpress.com
http://www.witpress.com

British Library Cataloguing-in-Publication Data

A Catalogue record for this book is available
from the British Library

ISBN: 978-1-78466-405-3
eISBN: 978-1-78466-406-0
ISSN: 1746-448X (print)
ISSN: 1743-3541 (on-line)

The texts of the papers in this volume were set individually by the authors or under their supervision. Only minor corrections to the text may have been carried out by the publisher.

Preface

This book contains the proceedings of the 10th International Conference on Waste Management and the Environment (WM20), which was held from 7 to 9 July, 2020. The conference was jointly organized by Wessex Institute (UK), University of Naples Parthenope (Italy) and Nagoya University (Japan), and we planned this meeting initially to take place at Madrid, Spain. However, the coronavirus pandemic did not allow us to open it as scheduled, and the conference committee decided to host the conference online.

Waste Management Conferences have been organized every two years since 2002; i.e., the first conference founded by Professor Carlos A. Brebbia (Wessex Institute) in Cadiz (2002), then held in Rhodes (2004), Malta (2006), Granada (2008), Tallinn (2010), the New Forest, home of the Wessex Institute (2012), Ancona (2014), Valencia (2016), and Seville (2018). A long-term accumulation of the data and discussion on waste management and the environment in the past two decades could contribute to the progress of waste management technology and its application to the social and economic system by international collaboration among researchers, engineers, companies and national/local governments.

The topics of the conference varied widely from very specific issues in local areas concerning the sanitary safety of municipal and hazardous wastes, and the development of each recycling technology, to large-scale environmental problems with deep concerns, e.g., global warming, climatic change, marine pollution and sustainable management of energy and material resources. It should be noted that the public consensus formation methodology has been developed among stakeholders including citizens, governments and companies. The main agendas of this online conference in WM20 were (1) municipal waste management, (2) landfill design and management, (3) reduce, reuse, recycle and recovery, (4) energy from waste, (5) hazardous waste, and (6) waste treatment technology, innovation and planning.

Even under such exceptional circumstances caused by the COVID-19 crisis, a good number of quality papers were submitted and accepted for publication, which is a testament to the dedication of academics and researchers in this field. Delegates were encouraged to access to the online conference website and download the papers and presentations in any order and as many times on demand, after that they could contact each other and promote discussion freely.

We are aware of the importance that the scientific community keeps on working in their institutions, then presenting their outcomes in the relevant academic societies and finally publishing their own research in the authorized books or journals. We have decided that the accepted papers should be published in a volume of the WIT Transactions on Ecology and the Environment, which are indexed in Scopus.

The Organizing Committee, who edited the proceedings, wishes to acknowledge the authors of the contributions, and the members of the International Scientific Advisory Committee who have peer-reviewed the submitted papers. The Editors would also like to express sincere thanks to Ms Priscilla Cook, the Conference Coordinator, and Ms Isabelle Rham, the Production Manager of WIT Press, in order to create this book.

The Editors hope this volume will open new horizon for the next Waste Management Conference after the current coronavirus crisis.

The Editors, 2020

Contents

BIODEGRADABLE PLASTICS FRAGMENTATION IN SOIL AND WATER: LESSONS LEARNT AND COMPARATIVE ASSESSMENT WITH HYDRO-BIODEGRADABLES

SULTAN MAJED AL-SALEM
Environment and Life Sciences Research Centre, Kuwait Institute for Scientific Research, Kuwait

ABSTRACT

In the current climate, there is a surge of controversy surrounding the topic of plastics. Undeniably and unavoidably, plastic has become a crucial part of this generation. However, the inappropriate disposal of plastic waste, as well as the increase in consumption, provokes serious environmental issues. The rising environmental awareness on a global scale has imposed considerable pressure for associated authorities to take actions. This includes the research into alternative options to conventional plastics which is latched on to a negative reputation. Nevertheless, the greenness of these so called environmentally friendly alternatives is often questionable. Since April 2015, the European Parliament and the Council established Directive (EU) 2015/720, has amended Directive 94/62/EC with regards to reducing the consumption of lightweight plastic carrier bags. Subsequently, the directive aims to reduce the level of littering of these bags which accumulates in the environment where more plastic-related problems (i.e. marine pollution) aggravates. In this work, a comprehensive fragmentation analysis was conducted on biodegradable plastic bags claimed to be green and environmentally friendly for arid environments. Various biodegradable compounds were tested and verified for such claims in this work. The first stage of the experimental campaign involved conditioning the samples under UV (to trigger decomposition) until fragmentation is reached; and record the Fourier Transform Infrared Spectroscopy spectra every 72 hours. The use of a reference material for comparative analysis was also applied, and the material was chosen to be polylactic acid. Furthermore, fragmentation in water was conducted where the plastic is conditioned in an aqueous environment. Lastly, the samples were mixed with sand and mud, and weathering induced fragmentation was continued. Infrared analysis was undertaken and the rate of biodegradation was determined. The above analysis was also used to draw lessons learnt from this exercise for oxo and hydro-biodegradables; and how to best utilize them as a waste management mitigation strategy.
Keywords: biodegradation, polyester, starch, polymer waste, waste management.

1 INTRODUCTION

Landfilling is considered to be the most common waste disposal method namely for municipal solid waste (MSW). With all the advances in technology available to it, the United States (US) generates some 267 million tonnes of MSW on an annual basis amounting to 4.5lb per person per day [1]. The total of 52% (139 million tonnes) is also landfilled on an annual basis. A major element of which are plastics. Plastic solid waste (PSW) comprises 12% of the global MSW composition [2]. Some reports show, namely for developing countries, that PSW comprise over 18% of MSW that is landfilled [3], [4]. Problematically, PSW is lightweight and inert on a chemical level. These might be some of its advantages when production and conversion takes place, however, are a major obstacle to landfill authorities when PSW is disposed of. Both film and rigid PSW also occupy valuable landfill space and has been related to various environmental stressors in the past [5]. A prime solution as of late is biodegradable plastics that have been popularised recently due to various advantages namely weight and size reduction of material with exposure to humidity and UV light [6]–[9]. Biodegradation refers to a degradation mechanism which

WIT Transactions on Ecology and the Environment, Vol 247, © 2020 WIT Press
www.witpress.com, ISSN 1743-3541 (on-line)
doi:10.2495/WM200011

involves the breakdown of organic chemical compounds via biological activity into more environmentally acceptable substances such as water, CO_2, new biomass or mineral salts of any present elements known as mineralisation. More specifically, the biological activity may be enzymatic, microbial and/or fungal activity. Furthermore, this process may take place in an oxygen-rich environment (aerobic condition) or in the absence of oxygen (anaerobic condition). In order for a plastic to be certified as a "biodegradable plastic", it must oblige to three specified parameters: (1) degree of degradation, (2) timespan of degradation, and (3) surrounding conditions of degradation. In addition, the certification bodies that are authorised to award such certifications, must follow standardised test methods (maybe more than one) to ensure the material satisfies all the appropriate specifications. It must be noted that all compostable materials are biodegradable, but a product that is labelled as biodegradable is not necessarily compostable. This statement is to clarify any misconception and prevent public from misleading information. The reader is referred to Al-Salem et al. [7]–[9] for the recent advances in standardization within the State of Kuwait for oxo-biodegradable plastics.

Biodegradable plastics can be classified into two main types, oxo-biodegradables and hydro-biodegradables. Oxo-degradable plastics are additivated with inorganic metal salts or pro-degradants. Whereas, a smaller portion of the group uses organic additives which are claimed to be consumed by micro-organisms. During the consumption process, the micro-organisms excrete acidic and enzymic substances which aid the breakdown of plastic material. Subsequently, the broken down and smaller pieces of plastic can be consumed by microbes with ease. This group is known as the "enzyme-mediated degradable plastics" [10]. On the other hand, hydro-biodegradables are a class of plastics with a portion of a renewable (naturally biobased) component such as rice or potato starch additives. The US guideline ASTM D6954 was once the sole available guide on testing oxo-degradable plastics for many years. As of 2009, increasingly number of guides and standards were developed throughout Europe and Middle East. Moreover, it must be acknowledged that guidelines and standards are separate entities. To elaborate, guidelines prescribe the procedures of different test methods, in which the ASTM D 6954 and BS 8472 are comparable in this case. However, standards consist of a higher level of details and quality which includes pass/fail criteria alongside the timescales in which the criteria must be satisfied by. For example, UAE/S 5009 and SPCR 141 are qualified as standards. In summary, the claim of conformity is not possible when testing complies solely to guidelines but without standards. According to a study published by the European Commission in 2018 [11], evidence shows suggests that oxo-degradable plastic is not suitable for any form of anaerobic digestion or composting. Thus, it will not fulfil the existing standards for packaging recoverable through composting within the EU. The Oxo-Biodegradable Plastic Association (OPA) stated that oxo-degradable plastics are not marketed for composting [12]. Dissimilarly to biodegradable plastics, despite some claims, oxo-degradable plastics are not compostable and will not degrade in landfill conditions. This is agreed upon by a large share of the industry. Furthermore, oxo-degradable plastics do not fulfil the specified requirements of (industrial and domestic) composting laid out by many accredited standards. This means that this type of material is not appropriate for compostable food service applications as an example which alternatively is a major share of solid waste landfilled.

In 2015, France was the first country to take concrete action to prohibit the production, distribution, sale, provision and utilisation of packaging or bags made either partially or completely of oxo-degradable plastics [13]. Recently, Spain became the second country to implement a similar law [14]. It is highly likely that other (EU) countries will follow suit as well. At the beginning of 2018, the European Commission (EC) has issued a statement that, the commission will take actions to "restrict the use" of oxo-degradable plastic across the EU member states. This statement was motivated by the concern that oxo-degradable plastics fragments into *microplastic* (MP) which subsequently pollute fauna and flora. This statement was published in the EC's plastics strategy on 16 Jan. 2018 [15]. A summary of standardisation and standards used in oxo-biodegradation is available elsewhere for the reader's consideration on the subject matter [8]. The commission further adds that there is no evidence to prove that oxo-degradable plastics offer any advantages over conventional plastics (which can be recycled).

Furthermore, the commission has revealed plans to regulate that all plastic packaging to be recyclable by 2030. The statistics is manly influenced by policies and legislations which are dependent on the country and regions. For instance, countries such as Pakistan, UAE and Yemen have legislations in favour of oxo-degradable plastic. Thereby, even for European based additive manufacturers, the main supply demands come from other regions. The estimation is determined on the basis of aggregated tonnage data obtained from European and non-European manufacturers. Nevertheless, from the available data, the Middle East and Central and South America were the two largest markets for pro-oxidant additives, each dominating around a third of the market. Recently efforts have taken place within the State of Kuwait to standardise the use of such materials with the aim of reducing waste accumulation [16]–[19]. Therefore, and as a continuing effort led by the Kuwait Institute for Scientific Research (KISR), standardisation of the use of hydro-biodegradable plastics in context of past effort with oxo-biodegradables is initiated. This communication represents the cornerstone of the aforementioned project where the experiences gathered from past efforts on oxo-biodegradables are compared with those from our ongoing research works on hydro-biodegradable polymer blends. The behaviour of the prior is investigated in various media to mimic their behaviour in various environmental sinks.

2 EXPERIMENTAL

The samples acquired were presumed to be environmentally friendly and biodegradable, which were secured from various sources within the state. The samples were cut into standard testing dimension specimens accepted by international protocols for thin plastic film characterization as described previously. For further details on types of the three bags which will termed Types 1 (yellow), 2 (blue), and 3 (white) herein, the readers are referred to our earlier work cited in this communication [7]–[9] including metal/additive content and experimental campaign for field testing. For the sake of comparison, plastic grades were purchased in the form of a standard grade of polylactide acid (PLA) which is a natural biodegradable material (Ingeo™ Biopolymer 4043D), as well as, a standard type of polyethylene (PE) bag used in an internationally recognised commercial outlet. Weathering in accelerated mode was conducted using stripes from the three different bags placed in a UV weatherometer machine from Q-SUN (Q-SUN Xe-3 Xenon Arc Test Chamber). The Q-SUN Xe-3 xenon arc chamber reproduces the damage caused by full-spectrum sunlight and rain. In a few days or weeks, the Q-SUN tester can reproduce the damage that occurs over months or years outdoors. The Q-SUN Xe-3 tester is a full-featured lightfastness, colourfastness, and photostability chamber. For conditioning the principles of ISO 4892-1/2016 – Part 1; were followed. To simulate the conditions in the Middle East and at the same time to speed up the fragmentation process we have used the following UV weathering parameters: Weatherometer: Eriskay, X3HS, T178, Filters: Daylight Q Filters, T211-Irradiance (nm): 340 (T169) – Power (Wm^{-2}): 0.35 – Chamber Temperature (°C): 43 – Black Panel Temperature (°C): 63 – Relative Humidity (%RH): N/A – Light Cycle only –

No spray cycles – Calibration Radiometer: CR20/340-D (Daylight filters) M142R. Specimens from the Types 1, 2, 3, commercial bag and PLA samples were placed for UV weathering and tested every 48 hours till 576 hours for carbonyl growth via the FTIR and every 100 hours till 200 hours for tensile performance. For Reference, 210 hours of weathering corresponds to approximately 2 to 3 months in real exposure time in sun. An IS10 Thermo-Nicolet FTIR machine equipped with an ATR attachment was used for the analysis of the bags. Sample were tested without any prior sample preparations. 32 scans per specimen were taken for better accuracy. For the carbonyl-oxidation analysis the weathered samples were analysed with the FTIR IS10. The peak areas from the 2,800 to 2,870 cm^{-1} and from the 1,600 to 1,850 regions were used for the carbonyl index (CI). An example of the analysis is shown in Fig. 1. See FTIR Index section for further details. In addition to the carbonyl, vinyl index was also calculated using the IR peak at 909 cm^{-1} and the reference peak 2,800 to 2,870 cm^{-1} (Figs 1 and 2).

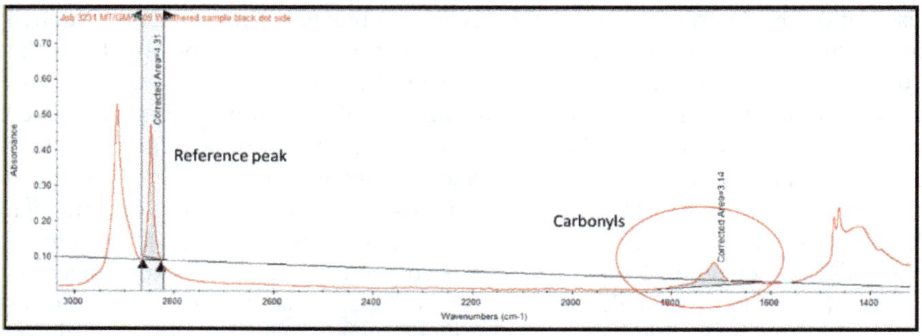

Figure 1: FTIR Spectrum highlighting the carbonyl formation due to degradation/ reference peak used in the analysis.

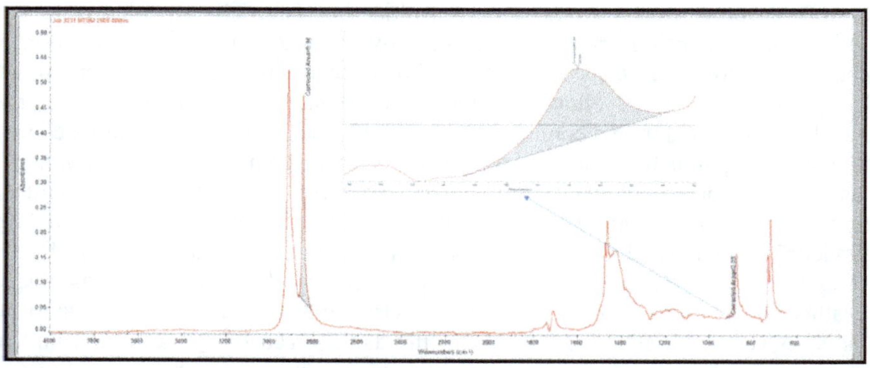

Figure 2: FTIR spectrum highlighting the carbonyl formation due to degradation.

Ageing in water took place by placing samples from the three different bags in glass bottles of water. The bottles with the samples were then placed in an air oven set at 65°C for ageing. Samples were inserted for testing the 22 Mar. 2018 and removed after 2 months.

3 RESULTS AND DISCUSSION

Specimens cut out from the commercial bags were placed in accelerated UV weathering chambers to assess the initiation and progression of the degradation process. The samples placed in the UV chamber were visually inspected on a periodical basis and tested under the FTIR for carbonyl growth, i.e. oxidation. In addition, the mechanical properties of the bags were determined upon UV exposure [17]. In the meantime, all samples were placed in water and at elevated temperature to investigate whether degradation is triggered in dark humid environments. Strips of the all the three samples were placed in water bottles at 65°C for ageing. A PLA sample along with a non-biodegradable bag form the commercial PE bag (Type 4) were also placed in water for testing. After almost 2 months of conditioning, none of the three types of bags showed evidence of degradation, including the supermarket bag (Type 4). But, the PLA film after approximately 20 days of conditioning became very brittle with a sugary texture when handled, indicating that the sample was degrading. Preliminary UV weathering tests revealed that the Type 2 blue sample started to fragment significantly after 65 hours, while the Type 1 and 3 bags were becoming brittle at approximately 200 hours. 200 hours of weathering corresponds to roughly 2 to 3 months in real exposure time in sun. Fig. 3 below shows the specimens after exposure to UV. This is a direct indication that the additives used to break down the PE bags are oxo-biodegradable and are activated with UV light [9] (Fig. 3). Composition of each bag is given elsewhere [7]–[9].

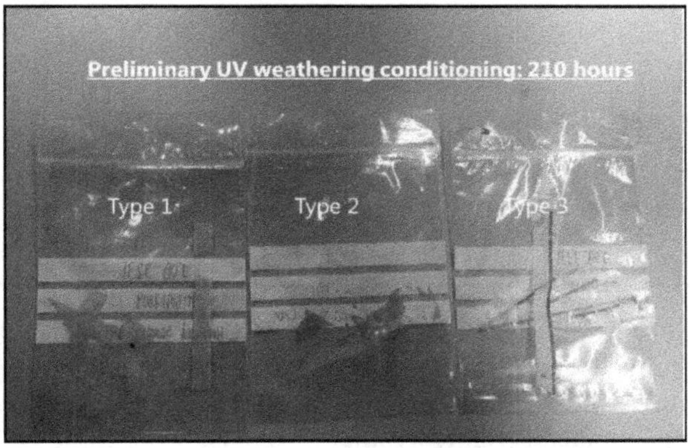

Figure 3: Preliminary UV weathering resistance tests in the three types of bags (210 hours).

Upon completion of the preliminary tests, more samples were placed in the UV chamber for in depth comparison of the oxidation and degradation processes. Specimens were removed from the chamber for infrared testing every 48 hours. From the collected FTIR spectra, the carbonyl indexes were determined and plotted against the time of UV exposure for each sample. Table 1 below shows the carbonyl index data along with the peak area calculations from the FTIR spectra. Fig. 4 shows the relation between carbonyl index and weathering time in hours for all three types of bags along with the reference commercial-international supermarket bag up to 384 hours. The linear trend lines are also shown. The slopes (gradient) are also highlighted in Fig. 4 with the Type 2 sample showing the highest slope and the Type 3 and commercial bag (Type 4) the lowest. In addition, Fig. 5 presents the graphs for all specimens across the full accelerated UV weathering conditioning. From the graphs we found that Type 2, blue bag showed faster growth of carbonyls on the surface and subsequently faster oxidation compared to the rest of the samples. The slope (gradient) of the blue sample was almost double than the slope of the yellow (Type 1) sample and much higher than the rest samples. Type 1, yellow bag showed slightly faster and higher oxidation than the Type 3 bag and less than the Type 1. Finally, according to the graph in Fig. 4, the Type 3 white bag showed similar oxidation behaviour with the commercial PE non-biodegradable bag. According to the graphs in Fig. 5, the best fitting for the Carbonyl growth behaviour is the polynomial, as a result the oxidation process does not have a linear relation with the weathering time.

The FTIR spectra focusing on the region where the carbonyl growth takes place at approximately 1,720 cm^{-1} are shown in Fig. 5. Near the carbonyl peak, an additional peak at around 1,800 cm^{-1} was evident in all samples including the reference commercial PE bag. According to the IR spectra the size of this additional peak was reduced with increasing the weathering exposure and the carbonyl peak was growing. It is very likely that these additional peaks are associated with the presence of UV absorbers or

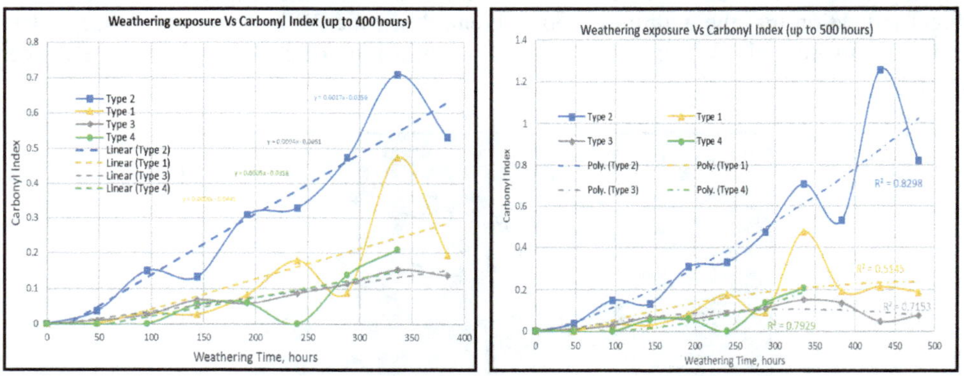

Figure 4: (Left) UV Weathering exposure vs. carbonyl index for all three types of bags plus the type 4 (commercial bag) reference PE bag, focusing on the up to 400 hours of weathering area. The linear fitting curves are also shown along with the slope gradient, (right) weathering exposure vs. carbonyl index for all three types of bags plus the reference pe bag, focusing on the up to 500 hours of weathering area. The polynomial fitting curves are also shown along with the square root of fitting.

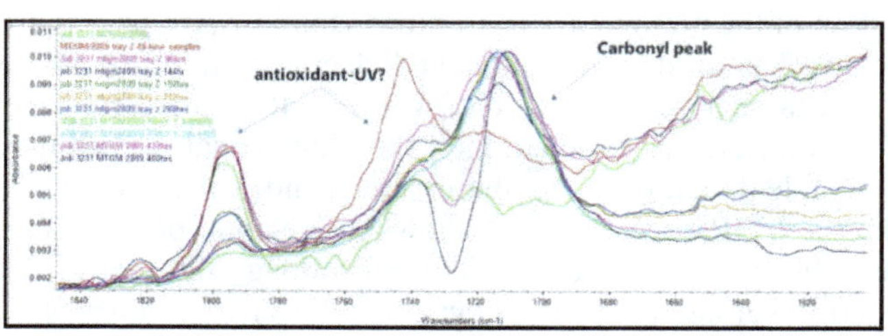

Figure 5: Carbonyl growth showing IR spectra for Type 1.

Table 1: Carbonyl index data for the investigated plastic films in this work.

	Type 1			Type 2			Type 3			Type 4		
Hours	1,800–1,650 cm^{-1}	2,865–2,790 cm^{-1}	CI	1,800–1,650 cm^{-1}	2,865–2,790 cm^{-1}	CI	1,800–1,650 cm^{-1}	2,865–2,790 cm^{-1}	CI	1,800–1,650 cm^{-1}	2,865–2,790 cm^{-1}	CI
0	0	5.27	0	0	5.7	0	0	5.62	0	0	3.7	0
48	0.19	5.58	0.03	0.17	5.9	0.03	0.11	5.75	0.02	11.25	1.64	0
96	0.26	5.44	0.05	1	6.02	0.17	0.25	4.92	0.05	8.95	2.06	0
144	0.21	5.69	0.04	1.05	6.02	0.17	0.36	5.23	0.07	0.29	3.44	0.06
192	0.59	5.86	0.10	1.66	5.71	0.29	0.36	4.94	0.07	0.22	3.39	0.06
240	1.38	5.44	0.25	1.93	5.66	0.34	0.6	5.11	0.12	7.49	1.12	0.00
288	0.62	5.61	0.11	2.87	5.71	0.50	0.83	5.63	0.15	0.42	2.98	0.14
336	2.21	5.44	0.41	3.99	5.96	0.67	0.86	5.87	0.15	N/A		
384	1.20	5.72	0.21	3.92	6.04	0.65	0.61	5.35	0.11			
432	1.14	5.1	0.22	4.32	4.11	1.05	0.60	5.71	0.11			
480	0.98	5.5	0.18	1.4	6.31	0.22	1.18	5.44	0.22			

antioxidants. But, as shown in Fig. 5, the Type 2, blue bag did not show a similar UV/antioxidant peak like the rest of the samples. It did show though a small peak at 1,660 cm^{-1} which could be related to antioxidants, but further analysis is required to support this argument. In any case, lack of antioxidants and or UV absorbers could be the reason why the Type 2, blue sample exhibited faster and bigger degradation-fragmentation than the rest of the samples. Furthermore, the vinyl index has been also calculated and plotted against the UV weathering hours. Fig. 6 presents the Vinyl index vs. UV weathering graph for the three types of bags including the reference commercial PE bag. From the graph it became evident that the reference bag from commercial PE bag (non-biodegradable) yielded much higher vinyl index during weathering exposure compared with the three types of bags. Also, the vinyl index of the Type 3 (white) bag did not change significantly during the UV weathering exposure.

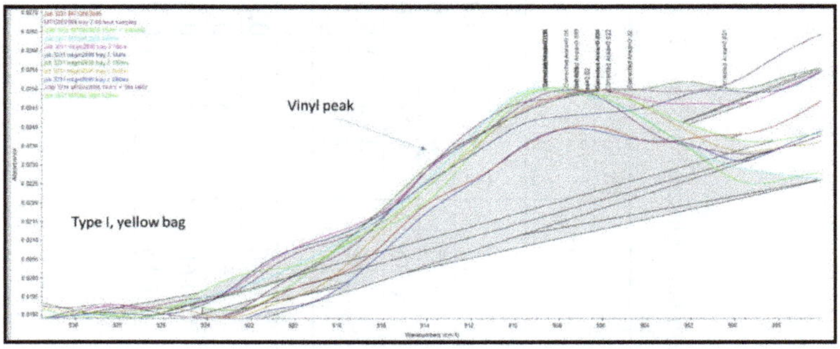

Figure 6: Type 1: Yellow sample IR spectra focusing on the vinyl growth.

Visual inspection of the weathered samples showed that the Type 1 yellow bag started fragmenting after approximately 288 hours, while the Type 2 blue sample after 192 hours. The Type 3 white bag did not show evidence of fragmentation before 400 hours. The reference clear bag from commercial PE bag showed traces of decomposition after 190 hours of weathering and PE specimens were blown away inside the UV chamber. The composition of each bag is given elsewhere [7]–[9].

The degradation process, simulating landfill conditions, will also be assessed and discussed in this section. The reference Type 4 was used for comparative assessment. The fragmented specimens from the analysis shown before will be mixed in wet sand and wet mud, compressed in individual boxes and was placed in the UV weatherometer for further conditioning. A wet cycle was added in the weathering program to keep the sand and mud samples wet in order to apply more “pressure” in the fragmented plastic pieces and accelerate the degradation process. Specimens were assessed periodically for degradation. For further UV weathering, the fragmented sample pieces were mixed with wet sand and wet mud and placed in rectangular plastic boxes of the same size. The sample preparation was kept consistent in order to apply packing, stacking pressure and environmental conditions similar to an actual landfill. Fig. 7 presents the filled boxes with all the samples mixed in sand. Both the boxes with the wet sand and the wet mud were then placed in a state-of-the-art accelerated UV (Xenon-arc) environmental chamber (Fig. 7). The environmental conditions were as per the following: RH: 60%, Temperature: 65°C using a daylight program. Sand and mud properties are depicted in Al-Salem et al. [19].

Figure 7: Samples mixed in mud and sand placed in the UV weathering chamber for further conditioning.

In total, the samples were exposed under UV for 1,200 hours which is more than a year in real time of sun exposure. The samples were mixed manually approximately every 72 hours. Every 200 hours (or 100 hours occasionally) of UV weathering a small amount of mixed sand and mud was removed from the boxes for degradation analysis. The degradation analysis was carried out visually, with GCMS and with FTIR, carbonyls. The presence of relatively large fragments in the mud and mud made the GCMS method very difficult to get meaningful results. The readers are referred to Al-Salem et al. [19] for further details. Following the completion of the UV weathering cycle, the sand and mud boxes were visually inspected for any remains of PE fragments. Fig. 8 shows an example of the sand and mud boxes exposed in UV for Type 1. The red circles in Fig. 8 highlight the presence of plastic fragments at all sand and mud samples. This is a clear indication that none of the bags were fully decomposed even after approximately 1 year of exposure in daylight (1,200 hours under UV). However, we have noticed visually the presence of much smaller polymer fragments in the mud samples.

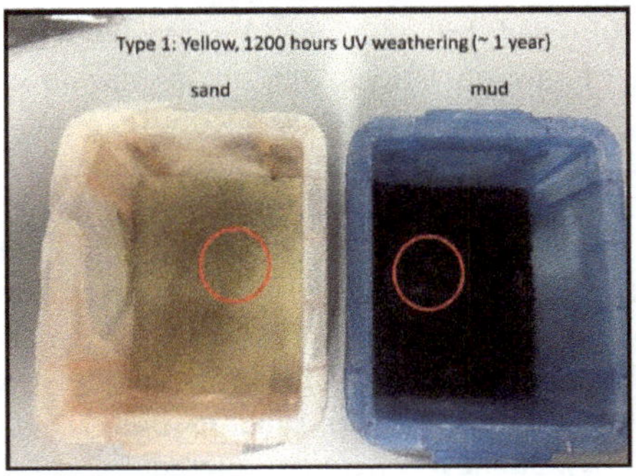

Figure 8: Type 1: Yellow bags in sand and mud, conditioned under UV for 1,200 hours.

The fragmented polymer pieces left in the sand and mud boxes after UV weathering, were collected and analysed with FTIR in order to assess the levels of carbonyls. The analysis was carried out and continued with the same method as previously mentioned. From the analysis, it is concluded that the Type 2 blue sample reached the maximum of oxidation between 400 and 500 hours i.e. no further oxidation-degradation took place in the sand box. In regards to the Type 1 and Type 3 samples, these have shown further degradation-oxidation during conditioning in the sand box. The fragments from the PE bag yielded higher carbonyl growth-oxidation in line with the Type 1 and Type 3 bags, with a slight decrease after 1100 hours. Finally, the oxidation behaviour of all the samples conditioned in sand and mud under UV was also studied. Fig. 9 shows the carbonyl index at 1200 hours of accelerated UV weathering vs. the mud and sand.

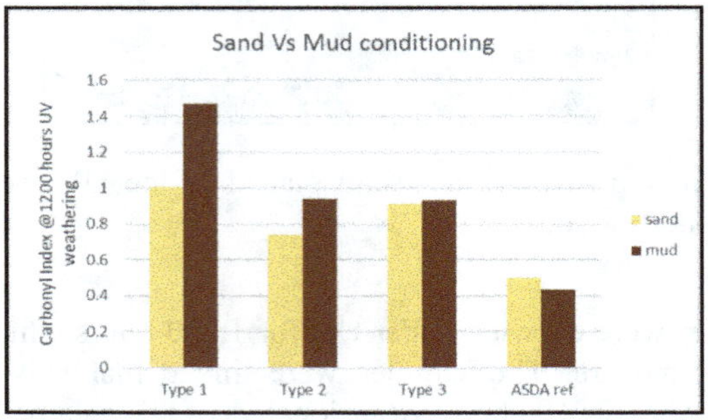

Figure 9: Carbonyl index at 1,200 hours of UV weathering of the samples conditioned in sand and mud.

From Fig. 9, it becomes evident the biodegradable samples conditioned in mud showed higher oxidation compared with the samples mixed with sand. This is not the case for Type 4. These results are in line with the visual observations undertaken and presented in previous sections. Breslin [20] studied the extent of deterioration of starch-plastic composites over a 2-year period for samples buried in a landfill. The limited degradation observed was attributed to the ineffectiveness of the pro-oxidant additive to catalyse the thermal oxidation of the plastic component of the starch-plastic composite under the environmental conditions present within the landfill. Song et al. [21] performed UV accelerated weathering followed by mechanical abrasion (MA) on various plastics to simulate the mechanism of microplastic evolution in marine environments. The rate of fragmentation by UV exposure duration increased more for PP than PE. These findings also support our current work, where it is suspected that the content of plasticizers, pro-oxidants and pro-degradants, will impact the fate of the biodegradable plastic exposed to open environments. It is also essential to keep track of potential chemicals that might pollute soils that plastics deteriorate in [22]. Viera et al. [23] assessed the impact of false claims by manufacturers in support of certain products which are marketed as biodegradables. Recently, Abed et al. [24] showed that oxo-biodegradable polymers are susceptible to bio-fouling. The formation of hydroxyl groups and carbonyl bonds, by Fourier-transform infrared spectroscopy (FTIR), suggests chemical degradation of the oxo-PE grade tested in the Arabian Gulf sea body.

4 CONCLUSION AND IMPLICATIONS ON HYDRO-BIODEGRADABLES

Accelerated UV exposure tests, revealed that all three types of bags started degrading after a certain time of conditioning. Upon 210 hours of exposure in UV, type 2 bags showed higher extent of fragmentation than the rest of the bags. 200 hours of UV exposure is approximately 2 months of exposure in UV in real conditions. From the analysis it was noted that Type 1 bag showed evidence of fragmentation after approximately 280 hours or 3 months in real time exposure to sun. The slope (oxidation rate) was lower than the Type 2, blue bag and higher than the Type 3 and the reference Type 4 bag. The presence of antioxidants appeared to have slightly delayed the carbonyl growth and oxidation. After approximately 5 months of simulated exposure, the specimens showed extensive brittle behaviour and fell apart when handled with hands. When compared with the reference PE bag from Type 4, Type 1 sample showed faster growth of carbonyl but similar fragmentation process. Type 3 bag: Type 3 white bag, showed traces of fragmentation after approximately 4 months of real time exposure. The presence of antioxidants might have hindered the oxidation and fragmentation process. The tensile strength of the White bag has been dropped by 43% after 200 hours of conditioning in the UV chamber, indication that the bag has not yet been fully degraded [19]. All three types of PE bags along with the reference non-biodegradable PE bag were placed in an accelerated UV chamber in humid, high temperature and daylight conditions to assess their total decomposition behaviour. To better simulate the landfill environmental conditions all samples were mixed with wet sand and mud and placed in plastic boxes (compacted) before placed into the UV chambers. The total time of exposure under UV for all specimens was approximately 1,200 hours. This corresponds to more than 1 year of actual exposure under the daylight (approximately 15 months). Sand and mud samples with the PE films were taken out from the chamber every 200 hours for visual and infrared inspection. FTIR, revealed the further oxidation of the Type 1, Type 3 and the reference Type 4 samples. Following the full weathering cycle, a series of fragments from all samples were evident still in the mud and the sand boxes, clear indication that the samples have not been fully decomposed. Quantitatively, more fragments were evident in the sand boxes compared with the mud ones. Also, the extent of oxidation at 1200 hours of UV was great in the mud boxes for the Types 1, 2 and 3 samples compared with the oxidation in the sand. In conclusion, it became evident from the results that the PE bags (Type 1, Type 2 and Type 3) did not decompose following a prolonged exposure under UV in a high temperature humid environment and mixed with sand and mud. It is imperative to consider these findings in-light with naturally filled biodegradables with starch and other chemicals, namely hydro-biodegradables to identify the best practice in a certain region or a country.

ACKNOWLEDGEMENTS

The author would like to extend his gratitude for the Kuwait Institute for Scientific Research (KISR) for funding research project EM065K. The author would like to thank the Kuwait Foundation for the Advancement of Sciences (KFAS) for funding this scientific mission.

REFERENCES

[1] USEPA, National overview: Facts and figures on materials, wastes and recycling, 2017. www.epa.gov/facts-and-figures-about-materials-waste-and-recycling/national-overview-facts-and-figures-materials.

[2] Kaza, S., Yao, L., Bhada-Tata, P. & Van Woerden, F., *What a Waste 2.0: A Global Snapshot of Solid Waste Management to 2050*, World Bank: Washington, DC, 2018. DOI: 10.1596/978-1-4648-1329-0.

[3] Franklin Associates Ltd., Characterization of municipal solid waste in the United States, 1960 to 2000 (Update 1988). Final report prepared for U.S. Environmental Protection Agency, Prairie Village (KS), 1990.

[4] Al-Jarallah, R. & Aleisa, E., A baseline study for municipal solid waste characterization for the state of Kuwait. *Waste Management*, **34**, pp. 952–960, 2014.

[5] Briassoulis, D. & Dejean, C., Critical review of norms and standards for biodegradable agricultural plastics, Part I. Biodegradation in soil. *Journal of Polymers and the Environment*, **18**, pp. 384–400, 2010.

[6] Al-Salem, S.M., Influential parameters on natural weathering under harsh climatic conditions of mechanically recycled plastic film specimens. *Journal of Environmental Management*, **230**, pp. 355–365, 2019.

[7] Al-Salem, S.M. et al., Thermal response and degressive reaction study of oxo-biodegradable plastic products exposed to various degradation media. *International Journal of Polymer Science*, 9612813, 2019.

[8] Al-Salem, S.M., Sultan, H.H., Karam, H.J. & Al-Dhafeeri, A.T., Determination of biodegradation rate of commercial oxo-biodegradable polyethylene film products using ASTM D 5988. *Journal of Polymer Research*, **26**, p. 157, 2019.

[9] Al-Salem, S.M., Al-Hazza'a, A., Karam, H.J., Al-Wadi, M.H., Al-Dhafeeri, A.T. & Al-Rowaih, A.A., Insights into the evaluation of the abiotic and biotic degradation rate of commercial pro-oxidant filled polyethylene (PE) thin films. *Journal of Environmental Management*, **250**, 109475, 2019.

[10] New Plastics Economy Initiative of the Ellen MacArthur Foundation. https://ecostandard.org/wp-content/uploads/oxo-statement.pdf.

[11] European Commission, Final Report: Study to provide information supplementing the study on the impact of the use of "oxo degradable" plastic on the environment, 2017.

[12] Oxo-Biodegradable Plastics Association (OPA), Oxo-bio technology reduces plastic waste in the environment. www.biodeg.org/OPA%20response%20to%20Ellen%20 MacArthur%20report%20-%2011.1.18.pdf.

[13] Oxo Statement, New plastics economy, 2017. www.european-bioplastics.org/over-150-organisations-back-call-to-ban-oxo-degradable-plastic-packaging/.

[14] Oxo-degradable plastics increasingly under fire in Europe, 2017. www.european-bioplastics.org/oxo-degradable-plastics-increasingly-under-fire-in-europe/.

[15] Communication from the Commission to the European Parliament, the Council, the European Economic and Social Committee and the Committee of the Regions, 2018. https://ec.europa.eu/environment/circular-economy/pdf/plastics-strategy.pdf.

[16] Al-Salem, S. et al., Investigating biodegradable bags standards and properties under Kuwait's environmental conditions, volume I: Aerobic biodegradation and degressive media impact. Final Report, Project Code: EM097C, May, 2019.

[17] Al-Salem, S. et al., Investigating biodegradable bags standards and properties under Kuwait's environmental conditions, volume II: Natural and accelerated weathering studies. Final Report, Project Code: EM097C, May, 2019.

[18] Al-Salem, S. et al., Investigating biodegradable bags standards and properties under Kuwait's environmental conditions, volume III: Thermal degradation kinetics. Final Report, Project Code: EM097C, May, 2019.

[19] Al-Salem, S. et al., Investigating biodegradable bags standards and properties under Kuwait's environmental conditions, volume IV: Development of governing standards for the use of biodegradable plastic bags in the state of Kuwait. Final Report, Project Code: EM097C, May, 2019.

[20] Breslin, V.T., Degradation of starch-plastic composites in a municipal solid waste landfill. *Journal of Environmental Polymer Degradation*, **1**(2), 1993.

[21] Song, Y.K., Hong, S.H., Jang, M., Han, G.M., Jung, S.W. & Shim, W.J., Combined Effects of UV exposure duration and mechanical abrasion on microplastic fragmentation by polymer type. *Environ. Sci. Technol.*, **51**, pp. 4368–4376, 2017.
[22] Tudor, V.C., Mocuta, D.N., Teodorescu, R.F. & Smedescu, D.I., The issue of plastic and microplastic pollution in soil. *Materiale Plastice*, **56**(3), 2019.
[23] Viera, J.S.C., Marques, M.R.C., Nazareth, M.C., Jimenez, P.C. & Castro, I.B., On replacing single-use plastic with so-called biodegradable ones: The case with straws. *Environmental Science and Policy*, **106**, pp. 177–181, 2020.
[24] Abed, R.M.M., Muthukrishnan, T., Al Khaburi, M., Al-Senafi, F., Munam, A. & Mahmoud, H., Degradability and biofouling of oxo-biodegradable polyethylene in the planktonic and benthic zones of the Arabian Gulf. *Marine Pollution Bulletin*, **150**, 110639, 2020.

IMPACT OF GROSS DOMESTIC PRODUCT CHANGE ON MUNICIPAL SOLID WASTE GENERATION IN MAPUTO, MOZAMBIQUE

AMAD H. A. GANI[1], ANTÓNIO G. DIAS[2] & ANTÓNIO A. R. MONJANE[1]
[1]Universidade Pedagógica de Maputo, Faculdade de Ciências Naturais e Matemática, Mozambique
[2]Universidade do Porto, Faculdade de Ciências, Departamento de Geociências, Ambiente e Ordenamento do Território, Portugal

ABSTRACT

The generation of Municipal Solid Waste (MSW) in Maputo grew from 2007 to 2017. Annual generation of the MSW increased from 319,010 tons in 2007 to 433,985 tons in 2017 and waste disposal in the dumpsite increased from 127,385 to 365,000. These numbers indicate that in 2007, waste disposal in dumpsite it was 39.93% from the total generated and in 2017, it was 84.10% of the total generated. The objective of this paper is to evaluate the evolution of MSW generation in Maputo municipality as a function of population growth and the variation of Maputo City's Gross Domestic Product (GDP) from 2007 to 2017. GDP varied, population and generation of MSW grew at different rates. At this time, the economy slowed down, and GDP fell from 7.4% to 6.8% due to the country's financial crisis. Although the national GDP and, consequently, of Maputo city fell significantly during this period, waste generation continued to grow because of the continuous migration of people from the other provinces to the capital city of Maputo. According to INE [1], the population of Maputo city had the lowest national growth, increasing slightly from 1,111,638 inhabitants in 2007 to 1,120,867 inhabitants in 2017, representing a population growth of 0.8%. During this period, MSW generation increased by 26.50%. The average capitation rate increased slightly from 0.96 kg/(inhab.day) in 2007 to 1.06 kg/(inhab.day).

Keywords: generation, dumpsite, Maputo, Gross Domestic Product (GDP), Municipal Solid Waste (MSW), capitation rate.

1 INTRODUCTION

Solid waste is the set of materials with solid consistency that the holder intends or needs to discard and may include what is left of the raw materials after use and which cannot be considered as a by-product or product [2]. Levy and Cabeças [2], defined MSW as similar material in consistency and nature generated from service and commercial sectors. The Municipal Solid Waste Management Regulation (MSWMR) of Maputo City defines MSW as being originating from domestic and commercial activities of settlements. The city's growth, emerging neighborhoods without an organized urban structure, coupled with the influx of people from other provinces and insufficient Eco-points, pressures consumer demand, making the MSWM in Maputo a major challenge.

The increase in consumer culture, combined with the growth of the world [3] and urban [4], population, change the standard consumption due to the long period of absence of their homes during the day. These changes led to an inevitable change in the economic and environmental framework [3]. MSW generation in Maputo follows the consumerist trend caused by the country's economic growth. Oliveira et. al. [3], understands that young people with a high academic level and new entrepreneurs are beginning to gain awareness of environmental responsibilities in relation to the MSW management. This awareness is growing, and it is noted that more and more people are sorting out MSW before and after being disposed of in the municipal Hulene dumpsite, with earnings for their family incomes. Aderoju et. al. [5], understands that MSWM is not only the responsibility of the government,

WIT Transactions on Ecology and the Environment, Vol 247, © 2020 WIT Press
www.witpress.com, ISSN 1743-3541 (on-line)
doi:10.2495/WM200021

but also requires self-discipline, change in people's perception and attitude to build a sustainable environment.

The annual world average of MSW generation is, according to Mantovani et al. [6], 1.4 billion tons, resulting in a per capita average of 1.2 kg/(inhab.day) while Kasaki and Kawai [7], which analyzed capitation rates from 157 countries, noted that the lowest capitation rate in the world is from the Ghana with 0.09 kg/(inhab.day), the highest is from the Antigua and Barbuda with 5.50 kg/(inhab.day) and the annual world average of 0.94 kg/(inhab.day). Mozambique's average capitation rate is 0.14 kg/(inhab.day), being the second lowest rate in the world after Ghana [7]. By contrast, the average capitation rate of EU countries is 1.22 kg/(inhab.day). This study is intending to evaluate the MSW generation in Maputo municipality as a function of population growth and the variation of Maputo City's Gross Domestic Product (GDP) from 2007 to 2017.

2 STUDY AREA

Maputo city is located on the western shore of Maputo Bay in the far south of the country, near South Africa and Swaziland borders. The boundaries of the Maputo municipality lie between the latitudes 25° 49' 09" S and 26° 05' 23" S and the longitudes 33° 00' 00" E (on Inhaca island) and 32° 26 '15" E and at an average elevation of 47m. The municipality has an area of about 346.77 square kilometers and a population density of 35 inhab/km^2. About 60% of the municipality's works in the informal sector, with the remaining 40% of the population in the formal sector divided into fisheries, agriculture, manufacturing, tourism and services. Throughout the year, the temperature ranges from 16°C to 29°C and is rarely below 14°C and can often exceed 40°C in the rainy season.

Maputo municipality consists of 7 municipal districts, namely: KaMpfumo, KaNlhamankulo, KaMaxaquene, Kamavota, KaMubucuane, KaTembe and KaNyaka. The last two districts were not included in this study because of their geographical location. The KaTembe district is located on the southernmost shore of Maputo Bay while the KaNyaka district is an island, so that its waste does not reach to Municipal Hulene dumpsite.

Data published by INE [8], point to a decrease in Mozambique's per capita GDP from US$ 458 to US$ 453 between 2007 and 2017. Maputo city has only 4% of the country's population but contributes with 18.5% of national GDP. INE [8] points out the sectors of commerce, transport and communications and manufacturing industry as the most significant, contributing respectively, with 29.6%, 29.5% and 12.4% of national generation. Mozambique has 53 municipalities and Maputo is the largest municipality in the country.

Trade is characterized by two types of markets: formal, comprising the retail and wholesale network, and informal, comprising fixed and mobile retailers scattered throughout the city streets and through formal and informal markets. In the Infulene green belt, a variety of vegetables are produced. In many suburban neighborhoods, poultry, rabbits and pigs are raised.

In urban neighborhoods, sanitation is served by a conventional domestic and rainwater sewage drainage system built in the colonial era and in older and emerging suburban neighborhoods, consisting essentially of septic tanks, generally unconnected to the public sewage drainage system. In terms of environmental legislation, the municipality has legal and normative MSWM instruments in which each of them complements in the absence of a more appropriate policy, namely:

- Decree 94/2014, of December 31st: Regulation on Urban Waste Management) [9];
- Resolution No. 86/AM/2008, of May 22nd, Posture of Cleaning of Solid Urban Waste in the Municipality of Maputo [10]; and
- Solid Waste Management Master Plan (SWMMP) [11].

2.1 Case study

Hulene's dumpsite is the case study of this work. It is located about 7km from Maputo city center and with the coordinate of the 25° 54' 03" S and 32° 35' 55" E at the center point (Fig. 1). The dumpsite has been operated since 1972 and reached the end of its useful life in 2016 and extended in 2018 for another 10 years due to compaction work. The dumpsite occupies an area of 17 hectares and has a maximum accumulated waste height of about 15 m.

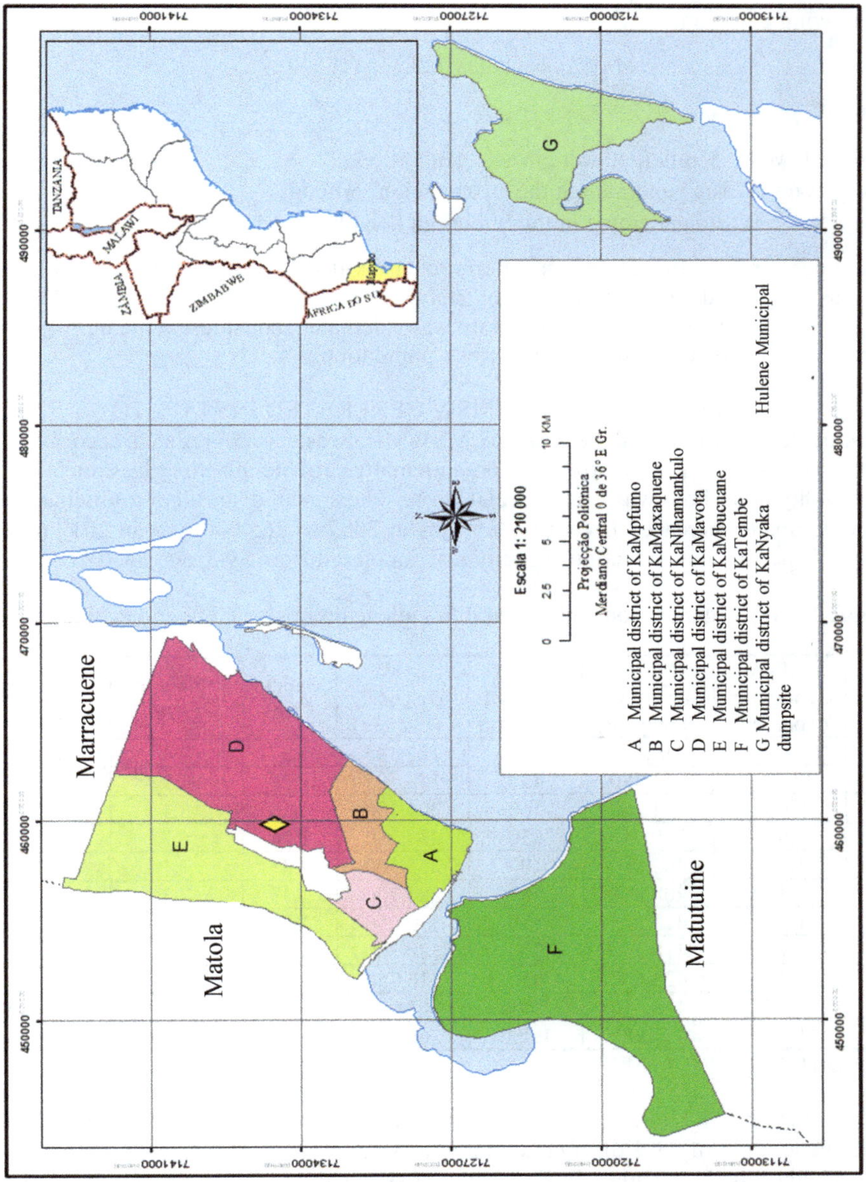

Figure 1: Location of Municipal Hulene dumpsite.

For the purposes of this study, MSW generation data were made public by the Maputo Municipal Council (MMC). GDP and population data are those officially published by the National Institute of Statistics (INE) of Mozambique in April 2019 and available on the internet and on the Government of Mozambique website.

The raw data of the paper/cardboard and disposable paper fractions were grouped and designated according to the urban solid waste segregation categories (Decree No. 94/2014), in paper/cardboard; the glasses and glass bottles were grouped into glasses; soft and hard plastics were grouped into plastics. The average solid waste generation was calculated using the following equation:

$$MSWg = (Wu \cdot 0.25) + (Ws \cdot 0.75),$$

where

- MSWg = Municipal Solid Waste generation,
- Wu = Waste generated in the Urban neighborhoods
- Ws = Waste generated in the Suburban neighborhoods

GDP per capita is the gross domestic product divided by midyear population. Maputo's GDP was calculated, considering the contribution of Maputo city indicated by INE [9], as 18.5% of national GDP. The capitation rate was calculated, considering the daily generation of MSW, divided by the number of the city's population.

3 COMPOSITION OF MSW GENERATED IN MAPUTO

According to Article 14 of the Maputo MSWMR, waste is segregated according to the following categories: paper/cardboard, organic matter, rubble, plastic, glass, metal, textiles, rubber, bulky household waste and special waste. The annual disposal of municipal waste in Hulene dumpsite increased from 127,385 tons in 2007 to 365,000 tons in 2017 (Table 1). During this period, the collection capacity also increased from 39.3% to 84.10%.

Table 1: MSW generation and disposal in Hulene dumpsite, GDP and capitation rate.

Year	*MSW Generation (Ton/day)	MSW Generation (Ton/year)	*MSW Disposal (Ton/day)	*MSW Disposal (Ton/year)	Maputo GDP	**National GDP	Capitation rate (kg/(inhab.dia))
2007	874	319,010	349	127,385	1.37	7.4	0.97
2008	972	354,780	437	159,505	1.28	6.9	0.99
2009	1,080	394,200	560	204,400	1.18	6.4	1.00
2010	1,093	398,945	650	237,250	1.24	6.7	1.01
2011	1,106	403,690	700	255,500	1.31	7.1	1.02
2012	1,120	408,800	780	284,700	1.33	7.2	1.03
2013	1,133	413,545	850	310,250	1.31	7.1	1.04
2014	1,146	418,290	900	328,500	1.37	7.4	1.04
2015	1,160	423,400	950	346,750	1.17	6.3	1.05
2016	1,174	428,510	1,000	365,000	1.11	6.0	1.06
2017	1,189	433,985	1,000	365,000	1.26	6.8	1.06

Source: Adapted from MMC*, INE**.

The suburban neighborhoods, with about 75% of the population, are responsible for the average generation of 73% of organic matter as shown in Fig. 2, against 64% generated in urban neighborhoods as shown in Fig. 3. Fig. 4 shows the MSW generation in Maputo city and it is observed that organic matter is the most commonly produced waste with an average of 71%.

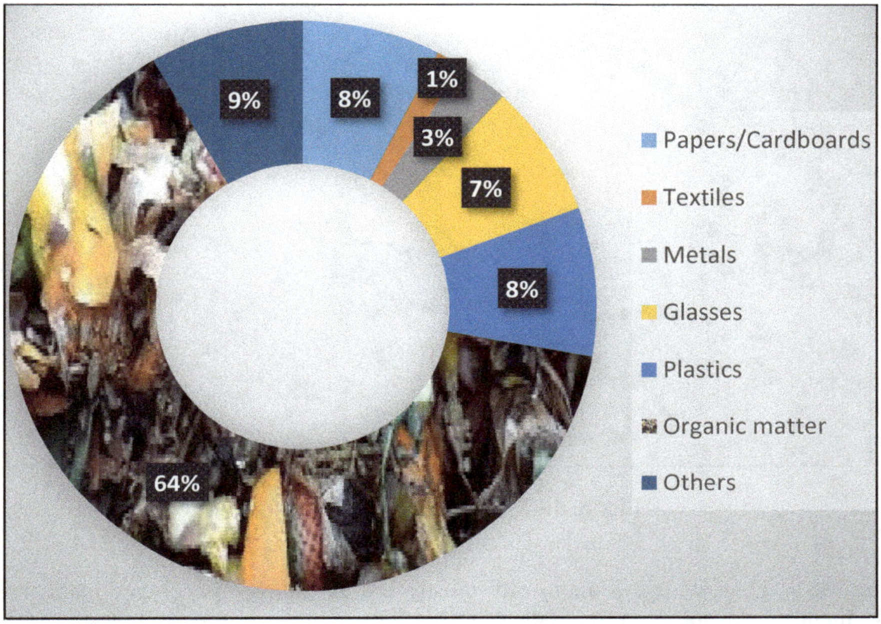

Figure 2: MSW generated in Urban neighborhoods.

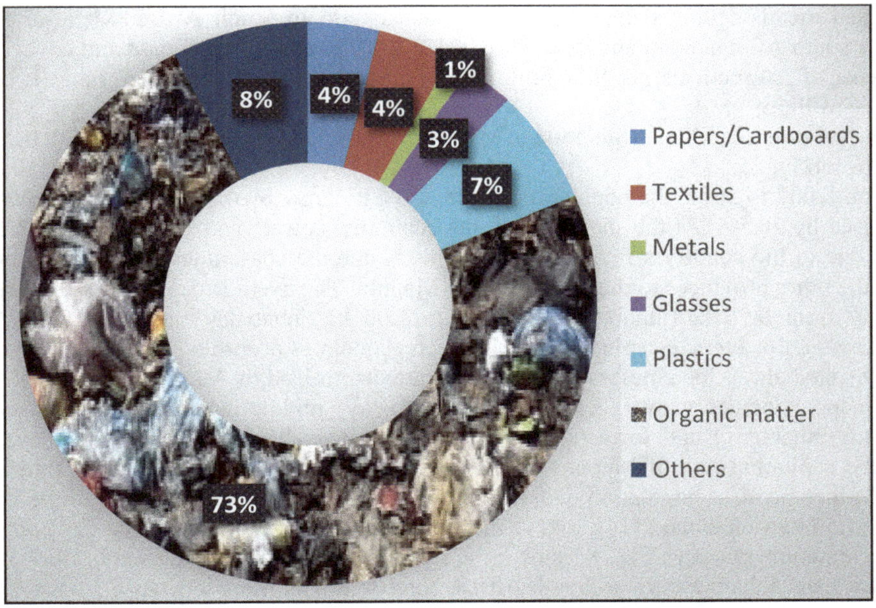

Figure 3: MSW generated in Suburban neighborhoods.

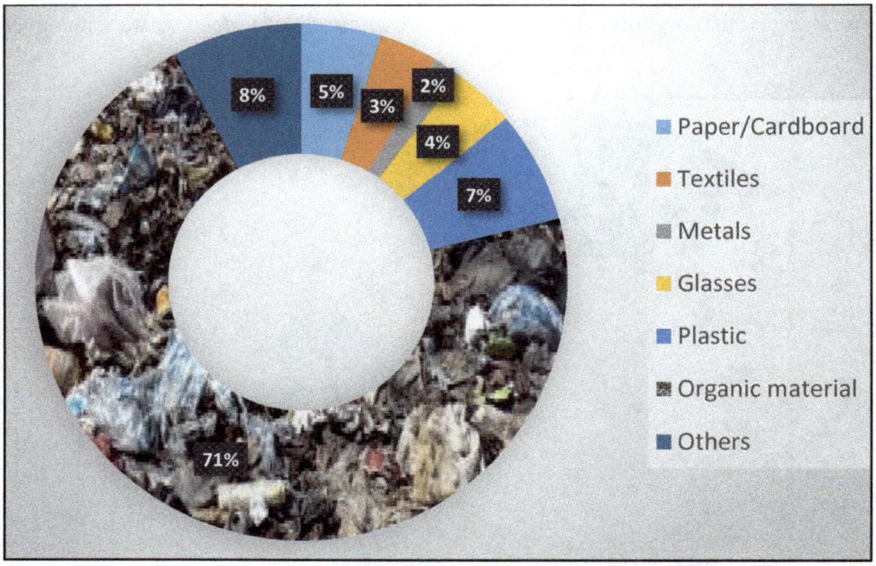

Figure 4: MSW generated in Maputo.

The MSW generated in Maputo city mostly are as follows: paper/cardboard, textiles, glass, metals, plastic, organic matter, and others. Bulky household waste, including construction waste, is used in foundations of new building under construction. The Hulene dumpsite does not receive this kind of waste.

Waste pickers essentially collect waste such as cardboard, plastics, pet and glass bottles, cans and metals selling them to companies that export to South Africa. Metal & Paper, Lda, a South African company, bought in 2018, 218,270 tons of cardboard and cover boxes at a cost of 3.0 meticais per kilogram, getting a yield of 654.810 million meticais (about US$9.6 million).

The evolution of MSW generation in Maputo over the years as a function of GDP behavior is shown in Fig. 5.

From 2007 to 2017, the population of Maputo City grew by 0.8% and MSW generation increased by 26.5%. At the same period, the economy slowed, and GDP fell from 7.4% to 6.8% due to the country's financial crisis. This is due the continuous migration of people from the other provinces to the capital city of Maputo. The average capitation rate increased slightly from 0.96 kg/(inhab.day) in 2007 to 1.06 kg/(inhab.day) in 2017 (Fig. 6). By comparison, the average capitation rate in green economies according to Elagroudy [12], is 1.2 kg/(inhab.day). Johannesburg's capitation rate determined by Masebinu et al. [13], for a high-income population averages is 1.91 kg/(inhab.day); middle-income class yields average 1.01 kg/(inhab.day) and lower-income class yields average is 0.92 kg/(inhab.day). The average capitation rate of Maputo city is equivalent to the average capitation rate of the middle-income class population of Johannesburg. The average capitation rate of the Lisbon city is 1.56 kg/(inhab.day) [14]. Fig. 7, show the MSW generation in Maputo city compared with capitation rate and Fig. 8, shows the average capitation rate of some cities in EU countries and Johannesburg, in South Africa, compared with Maputo city.

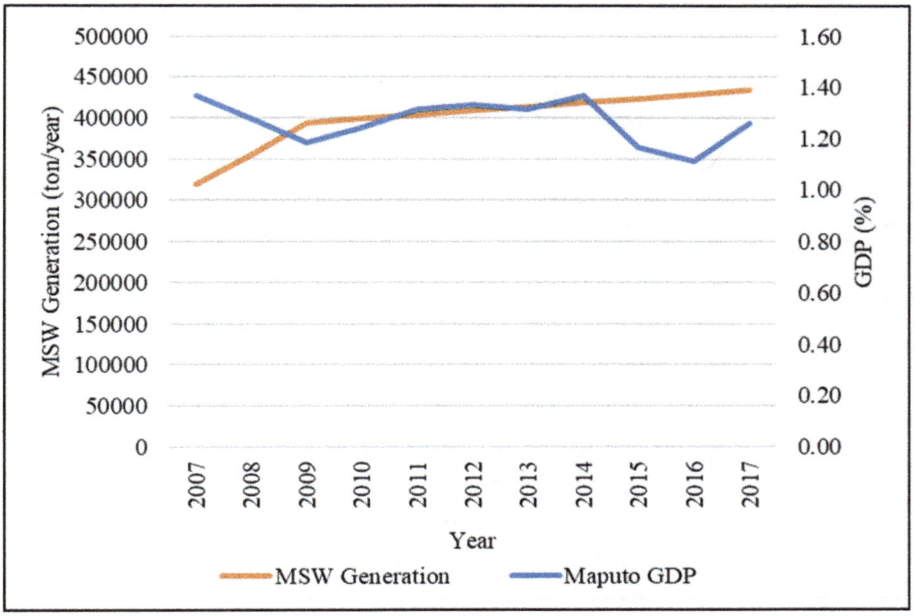

Figure 5: MSW Generation and GDP.

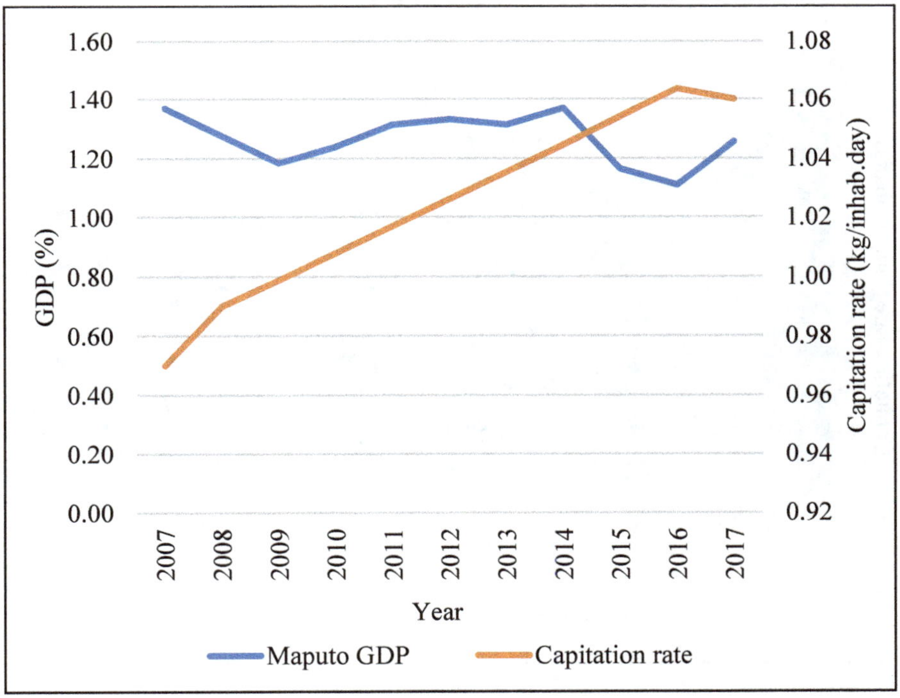

Figure 6: Capitation rate and GDP.

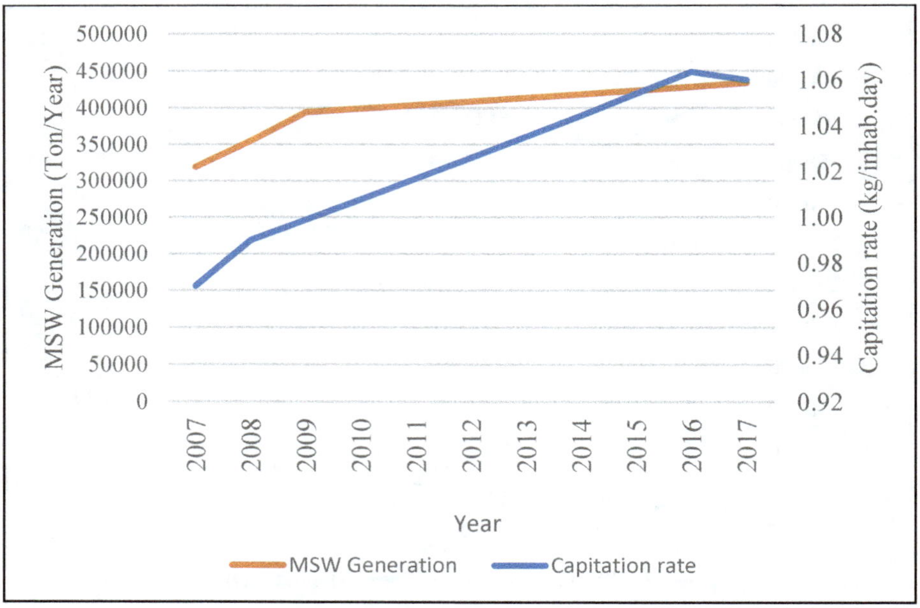

Figure 7: MSW Generation and Capitation rate.

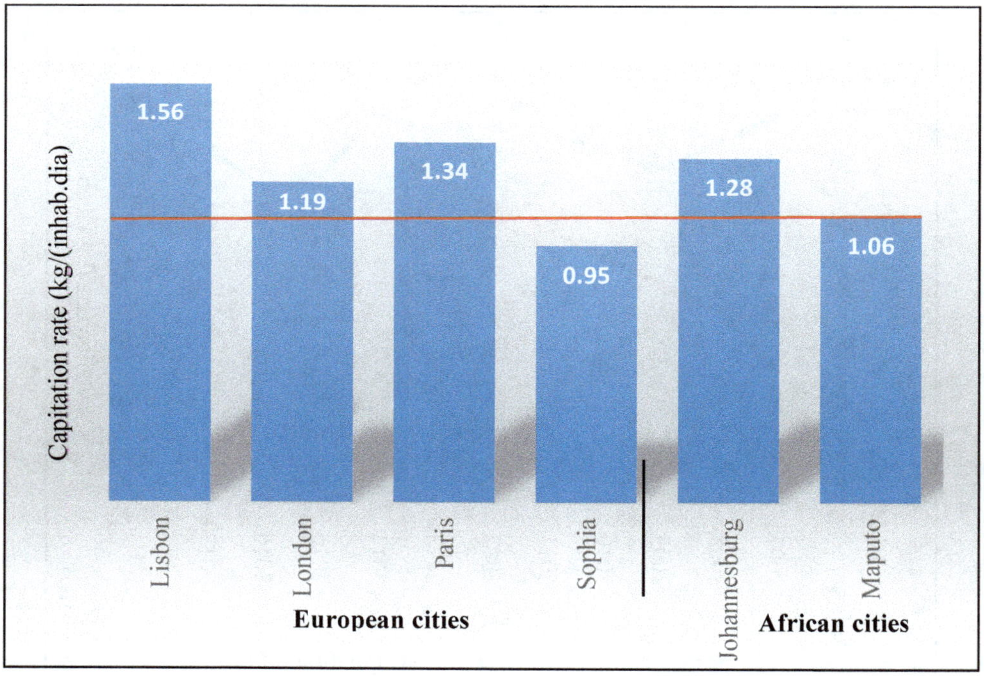

Figure 8: Capitation rates of the EU and African cities. *(Source: Adapted from Masebinu et al. 2017 [13] and Brussels DG ENV 2015 [14].)*

4 CONCLUSION

The city's population grew by 0.8% and MSW generation increased by 26.5%. Although national GDP fell from 7.4% in 2007 to 6.8% in 2017 with some variation, MSW productions continued to grow. The average capitation rate of Maputo city grew over the same period from 0.97 kg/(inhab.day) to 1.06 kg/(inhab.day) and is close to the average capitation rate of the middle-class population of Johannesburg. This rate is above the average capitation rate of the Sofia city and below the Lisbon, London and Paris cities and above of the average Mozambique capitation rate.

Organic matter produced in Maputo is the largest fraction of MSW, with a total of 71%, indicating higher demand for food compared to other products. Suburban neighborhoods with 75% of the population, produce 73% of organic matter against 64% of urban neighborhoods that have 25% of the city's population. In contrast, the production of paper/cardboard, textile, glass, metal, plastic, and other fractions in urban neighborhoods is 36%, while in suburban neighborhoods is 27%. Over the period studied, the effectiveness of the MSW collection system in Maputo City increased from 39.93% to 84.10%, representing a growth of 44.17% and a difference of 237,615 tons of waste disposed.

REFERENCES

[1] INE, Apresentação dos Resultados do IV Recenseamento Geral da População, Censo de 2017. Instituto Nacional de Estatística de Moçambique, Maputo 2019. www.ine.gov.mz.

[2] Levy, J.P. & Cabeças, A.J., *Resíduos Sólidos Urbanos – Princípios e Processos*, 1ª Edição, Editado por AEPSA: Lisboa, 2006. ISBN: 989-95059-0-0.

[3] Oliveira, F.F.S., Mendes, B. & Lapa, N., Resíduos. *Gestão, Tratamento e sua Problemática em Portugal*, LIDEL, Lisboa, p. 71, 2009. ISBN: 978-972-757-504-6.

[4] Yoshida, M. & Agency, C., *Situation of Municipal Solid Waste Management in African Cities – An Interpretation of the Information provided by the First ACCP Meeting*, 1(June), 2000–2015, 2018.

[5] Aderoju, O.M., Dias, A.G. & Guimarães, R., *Wastes: Solutions, Treatments and Oportunities*, eds, Vilarinho, Castelo & Russo, London, ISBN 978-1-138-02882-1, 2015.

[6] Mantovani, B., Cavion, G. & Finkler, R., *Geração de Resíduos Sólidos Urbanos e o Produto Interno Bruto (PIB): Análise da Realidade de Caxias do Sul – Rio Grande do Sul*, 2017.

[7] Kasaki, T. & Kawai, K., *Revisiting Estimates of Municipal Solid Waste Generation Per Capita and Their Reliability*, 3R Policy Maker, Japan, 2015.

[8] INE, PIB Provincial de 2007 a 2017, Maputo, 2019. www.ine.gov.mz/estatisticas/estatisticas-economicas/contas-nacionais/anuais-1/pib-provincial-2007-2013-2017-quadros/at_download/file.

[9] Decree No. 94/2014, Regulation on municipal waste management, Maputo, 31 December 2014.

[10] Resolução Resolution No. 86/AM/2008, Posture of cleaning of solid municipal waste in the Municipality of Maputo, Maputo, 22 May 2006.

[11] Maputo, C.M.C., Solid Waste Management Master Plan (SWMMP), Maputo, 2008. www.giz.de/en/downloads/giz2012-en-economic-instruments-mozambique.pdf.

[12] Elagroudy, S., Warith, M.A. & Zayat, M., *Municipal Solid Waste Management And Green Economy*, Global Young Academy Jägerstr: Berlin, Germany, 2016. ISBN: 978-3-939818-65-6. www.globalyoungacademy.net.

[13] Masebinu, S., Akinlabi, E.T., Muzenda, E., Aboyade, A.O., Mbohwa, C., Manyuchi, M. & Naidoo, P., a review on factors affecting municipal solid waste generation. Presented at: *2nd International Engineering Conference*, Federal University of Technology, Minna, Nigeria, 17–19 Oct. 2017.
[14] Brussels, DG ENV, Assessment of separate collection schemes in the 28 capitals of the EU. Final report, Copenhagen Resource Institute, BIPRO: Brussels, 2015. Reference: 070201/ENV/2014/691401/SFRA/A2.

WASTE VALUE POTENTIAL ANALYSIS OF MUNICIPAL SOLID WASTE PRODUCED IN THE PERI-URBAN AREA OF ZHAOQUANYING, CHINA

VICKY SHETTIGONDAHALLI EKANTHALU[1], SAFWAT HEMIDAT[1], SUSANNE HARTARD[2], GERT MORSCHECK[1], MONA-MARIA NARRA[1], SATYANARAYANA NARRA[1,3], JAN SPRAFKE[1] & MICHAEL NELLES[1,3]
[1]Faculty of Agriculture and Environmental Science, Universität Rostock, Germany
[2]Environmental Campus Birkenfeld, University of Applied Sciences Trier, Germany
[3] Deutsches Biomasseforschungszentrum (DBFZ), Germany

ABSTRACT

The recovering and recycling of municipal solid waste (MSW) plays a most crucial role in sustainable waste management and environmental preservation. The rapid urbanization and migration of people is urging the bigger cities of china to handle significantly higher amount of waste. Likewise, the overdependence of peri-urban regions like Zhaoquanying town (ZQY) on bigger cities is making the concept of recycling and recovery a major challenge and simultaneously causing a higher burden on urban areas for sustainable waste management. In this concern, it is crucial to make peri-urban regions like ZQY self-reliant in their waste management. The current study provides the integration of experimental analysis and theoretical evaluation to determine the waste value potential of the MSW produced in ZQY. The experimental sorting analysis of MSW in ZQY has revealed that the major component of MSW consists of recoverable waste with an average of 28% (kitchen and other organic waste) and recyclable waste with an average of 20% (paper, plastic, and metal). Despite the having a higher amount of recoverable and recyclable waste, all the produced waste in ZQY is currently ending up in the Shunyi central incineration plant without efficient waste value recovery. The experimental analysis from the organic waste in ZQY has proved to have a higher potential in producing better quality methane and compost. The analysed results reveal that organic waste produced in ZQY has comparatively higher value potential than the recyclable waste available in ZQY, and the total estimated value potential of municipal waste produced in ZQY is between 2.93–3.66 million Yuan/year. The outcome of this study will provide policymakers with the first-hand information and reasoning for directing efforts to achieve waste recovery and recycling in their current waste management practice in ZQY.

Keywords: municipal solid waste in China, waste management, waste value, anaerobic digestion, composting, waste recycling.

1 INTRODUCTION

The steady rise in the human population has resulted in the generation of a large quantity of waste. Especially developing countries like China have a relatively higher composition of bio-waste which together with the challenges (environmental, social, and economic) provides various opportunities for the production of value-added products. In China, the average per capita volume of municipal waste in cities is 1.17 kg/day, in townships 0.79 kg/day, and in villages 0.5 kg/day [1], [2]. According to the Statistical Yearbook of Urban and Rural Construction (Ministry of Housing and Urban-Rural Development of People's Republic of China, MOHURD 2020), 53% of China's municipal waste was disposed of in landfills in 2017. Another study shows that in 2014 only 10% of solid municipal waste generated in villages was properly treated [1]. Several developed countries such as Belgium, Denmark, Germany, Japan, and Sweden have a municipal waste reuse index of over 90% [3]. Bio-digestion, composting, incineration, and recycling are the most popular methods of treating organic and inorganic waste to produce useful products such as biogas (heat and

WIT Transactions on Ecology and the Environment, Vol 247, © 2020 WIT Press
www.witpress.com, ISSN 1743-3541 (on-line)
doi:10.2495/WM200031

electricity) and organic fertilizers. Similarly, recyclable waste can be recycled and sent back to the production chain as raw materials creating several economic, energetic, environmental, and social benefits [4].

This research proposes a scenario for treating and reutilising the municipal solid waste produced in Zhaoquanying (ZQY) by bio-digestion, composting, and recycling. The proposed scenario is based upon the household waste analysis, collection of additional data during the on-site visits and discussions, assessment of additional data in the literature comparison that was made within the consortium of the PERIVAL project. Recycling is proposed to understand the value potential for recyclable waste such as paper and cardboard, plastic, glass, and metal that is present in solid waste from ZQY. Among the various bio-waste treatment options, composting and biogas production has gained wide popularity owing to its associated benefits such as hygienization of waste, cost-effectiveness, and waste value creation [5], [6]. In this concern, experimental and theoretical analysis has been conducted on composting and biogas to understand the bio-waste value potential of ZQY.

The study results manifest the environmental benefits and socio-economic potential that can incur by sustainably treating the waste produced in ZQY. Several additional benefits include the reduction in the operational costs of solid waste management and a significant decrease in the overdependence of peri-urban areas (ZQY town) on bigger cities for their waste management. The treatment of organic waste by anaerobic digestion and composting enables in the production of additional energy and bio-fertilizer which can supplement the network of energy and fertilizer uses in the locals. In addition to the reductions in emission, decrease in the soil and water contamination amongst other potential positive outcomes.

2 MATERIAL AND METHODS

2.1 Study area

As part of the PERIVAL – Waste to value chains in peri-urban environments projects, ZQY was selected as a pilot town in the periphery of Beijing. ZQY town is a part of the Shunyi district's administrative divisions. ZQY town consists of 24 villages with 30,182 inhabitants and produces approximately 38 tons of municipal waste/day. ZQY was particularly selected taking into account of eco-industrial potential in intensive horticulture in the Beilangzhong district. The waste from 24 villages of the ZQY municipality is collected every day from the doorstep and transported to a central waste transfer station in ZQY and treated together. Delivered waste is treated together by compacting to remove leachate and transported further to a centralized waste incineration plant in Shunyi and incinerated leaving behind lots of recyclable and removable resources. ZQY has the optimal prerequisite to link the produced organic waste (biogas and/or heat, electricity, and fertilizer) to heat greenhouses for heated orchid growing in greenhouses and to fertilize the crops. The primary and practical aim of this research is to explore the waste value potential of the municipal waste produced in ZQY which can later be used as a model study for analysing waste value potential in typical Chinese peri-urban settlement areas. In this concern this research integrates the experimental and literature data to provide the detailed analysis results on biogas production, composting, and recycling.

2.2 Waste sampling procedure and sorting analysis

The sorting analysis was carried out on the mixed residual waste consisting of waste materials with high heterogeneity. The waste for sorting analysis was delivered from the central waste transfer station of ZQY before compaction. The waste sorting analysis nomenclatures were adapted from literature data [7], [8] and optimized to fit the local waste fraction behavior (see Fig. 1). The sorting analysis was carried out in an open area with the presence of a roof to restrain the addition of rainwater or moisture that can alter the characteristics of waste. To understand the seasonal waste composition changes, two separate waste sorting analysis was carried out comprising of one each in spring 2019 (27 March–1 April) and summer 2019 (5–11 July). The obtained waste samples were initially sieved using a 40 mm sieve and the coarse fraction with size > 40 mm was manually sorted by hand. During the sorting analysis, qualified samples were filled into paper bags and pre-dried using a makeshift ventilator system and transported to Germany to determine the physical and chemical properties of the collected sample. However, glass, metals, and mineral wastes have no heating value and are relatively inert, thus these waste streams were excluded from further chemical analysis. The representative samples of kitchen waste, other-organic waste, and fine fractions (< 40 mm) were finely grounded to make homogeneous samples of grain size 3 mm and were chemically analysed in Germany.

2.3 Biogas

The biogas analysis was carried out at the laboratory site of the University of Rostock equipped with 50 batch-fermenters. Each batch-fermenter is of 60 liters capacity and installed with the magnetic stirrer for continuous stirring. The tests are carried out in accordance with the German directive VDI 4630 at mesophilic temperature. Batch-tests are featured by one-time feeding of the substrate and the logging of the amount and quality of the biogas over the test period. Batch tests aids in analysing the biogas and methane yield of the tested substrate, anaerobic digestion rate, and inhibitory effects of the substrate within the tested range of concentration. According to the directive VDI 4630, the sample mass is determined by the ratio of dry organic matter of the tested substratum and the inoculum. Thus, the sample mass is about 0.2–0.3 kg of the substrate from ZQY is used with the sewage sludge from a municipal sewage water treatment plant as inoculum.

The produced biogas is stored in special gas sampling bags which are deflated regularly. While deflating, the volume and composition of the produced biogas are measured. The volume is measured by a RITTER drum-type gas meter. For the representation of the results, the gas volume is converted to the volume under normal conditions (273 K, 1013 hPa). For the measurement of the gas composition, an EHEIM gas analyzer with sensors for CH_4, CO_2, O_2, and H_2S is used. The duration of the batch test depends on the daily increase of the biogas volume. Tests are aborted if the daily increase is lower than 1%.

2.4 Composting and recycling

To assess the value potential of recyclables and compostable waste, an economic model is proposed. In this concern besides the population, accurate estimates of waste quantities generated and collected separately are required. Material that is already separated and diverted can be quantified through direct measurement, but in the case of organics mixed with other waste types, solid waste professionals must extrapolate quantities from known values that represent their percentage of the incoming waste stream. To assess the value potential of recyclables the waste fractions from PPC (paper, paperboard, and cardboard), glass, plastic packaging, and metal are considered. After calculating the actual weight of the sample concerning their dry matter content, the economic value potential is analysed based on the market value for the recyclables in China. Similarly, to understand the value potential of composting the waste fractions from kitchen waste, organic waste and < 40 mm fractions were used. The reason behind considering the waste fractions < 40 mm is that this fraction of waste had significantly larger amounts of compostable fractions. High levels of heavy metals represent an obvious concern when the compost is destined to be applied to food crops. Similarly, the concentration major elements such as nitrogen, phosphorus, and potassium in the produced organic waste play a major role in deciding the quality of the compost in China. In this concern, the selected samples were analysed for their heavy metal and nutrients concentrations.

3 RESULTS AND DISCUSSION

3.1 Waste sorting analysis

The waste composition of ZQY was dominated with a fraction < 40 mm size with a close following of the kitchen waste. Kitchen waste consisting of vegetable and animal-derived food waste constituted for 28.69% in spring and 18.45% in summer. Despite the fact that waste composition was dominated with a < 40 mm fraction, this particular fractions had a major composition of finer kitchen waste and a comparable portion of mineral and hygienic materials.

Fig. 1 depicts the waste composition in ZQY. The fine fraction (0–10 mm fraction) includes all kinds of materials present in the waste but represents mainly a mixture of mineral and organic particles. Food waste composition consisting of vegetable and animal-derived food waste constituted for 28.69% in spring and 18.45% in summer. Followed by food waste, plastic packaging waste constituted 10.69% and 15.39% of total waste generated in spring and summer, respectively. PPC had a composition of 2.65 and 2.62%, glass constitutes 2.28 and 3.74%, and metal with 1.85 and 1.01% respectively in spring and summer 2019. A relatively higher fraction of undefined waste was found in spring season 4.39% compared to summer 1.70%. The composition of textile waste was 3.06 and 4.49% and hazardous waste/e-waste were 0.55 and 0.27% in spring and summer season respectively. The percentage of total recyclable waste generated from PPC, glass, and metal constituted 6.78 and 7.37% in spring and summer, respectively. Except for PPC, the analysis results of recyclables were similar to the results obtained from most of the middle-income countries [9]–[11].

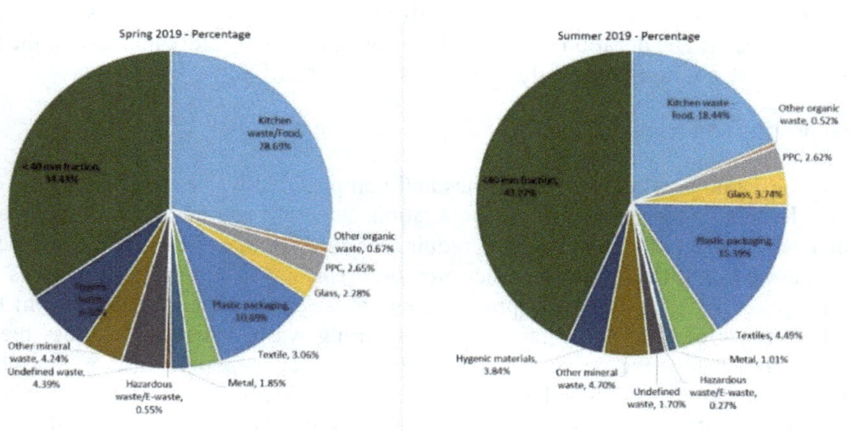

Figure 1: Composition of mixed solid waste in ZQY town in spring and summer 2019.

There is a significant decrease in the composition of kitchen waste in summer compared to winter. In contrast, the fractions < 40 mm in the summer season has increased in comparison with winter. As explained earlier, the fraction < 40 mm is dominated by the finer fractions from kitchen waste and the decrease in the kitchen waste composition in summer can be directly correlated with the increase in the composition of < 40 mm fractions. The composition of recyclable wastes is lower compared to the average MSW composition in the EU and reason seems to be the source-segregation of recyclables and informal waste collections in ZQY [12]. As explained earlier ZQY is the peri-urban city where most of the produced recyclable wastes are retained by households for further sale or collected by the informal sectors. As the characteristic of the organic waste does not provide the direct recycling/monitory benefit this category of waste ends up in the incineration plant without resource recovery. Variations in the source-segregation pattern in ZQY town may potentially explain differences in the composition of recyclables. Nevertheless, additional factors such as income levels, local sorting guidelines, consumption patterns, demographics, and development, in general, may influence the composition.

3.2 Biogas potential in ZQY

3.2.1 Organic residue available in ZQY for biogas production

The biogas plant in ZQY can play an important role in addressing the organic waste management and pollution control of agriculture residues. Introducing biogas production in ZQY waste management will aid in the efficient utilization of waste resource and promotes the development of the circular economy. ZQY produces nearly 38 t of MSW/day of which nearly 23.57% of waste was solely from food/kitchen waste (analysis made in spring and summer 2019). In addition to the organic waste from the household, there is a huge amount of biomass available in the form of agriculture residues in ZQY. Despite the fact of rapid urbanization due to the fact of being located in close proximity to Beijing, agriculture has still been of greater importance in ZQY. According to the Beijing Shunyi 2018 statistical yearbook; wheat, corn, and orchids are the most important crops that are grown in ZQY [13]. According to the literature, there were about 3218 tons of collectible agriculture residue available in ZQY in 2017 [13] (see Table 1). The available biomass in ZQY has significant potential to produce high calorific value methane. The produced methane from available biomass has the potential to reduce GHG emissions by replacing fossil fuels and biomass energy simultaneously providing sustainable waste management pathways.

3.2.2 Biogas and fermented residue resource potential in ZQY

Developing the biogas plant in ZQY can exploit the energy value present in the organic residues. Additionally, the utilization of the fermented residue after biogas production can not only help to generate additional revenue but also reduces the use of chemical fertilizer and aids in reducing GHG emissions. Considering all the available biomass in ZQY there is a possibility to set a biogas plant of 238 kW power (assumption includes 7,000 operation hours/annum and 30% of electrical and thermal efficiency). According to the experimental analysis there is a potential to produce 586,522.96 m^3 CH_4/a, which is equivalent to 1,500 MWh/a of electricity and 1,228 MWh/a of heat (see Table 1). Considering the selling price of electricity at 0.55 Yuan/kWh and heat at 0.16 Yuan/kWh [6] a revenue of 1 .02 million Yuan/year can be generated by selling heat and electricity produced by biomass in ZQY.

The anaerobic fermentation residue contains many major elements such as nitrogen, phosphorus, and potassium that are necessary for plant growth. Co-digestion of agricultural waste with the kitchen waste produced in ZQY can produce approximately 5,417.48 tons of fermented residue/a. Additionally, the economic feasibility of any biogas plant greatly depends on the marketability of the produced digestate. Co-digestion of agricultural waste with kitchen waste produced in ZQY can produce 5,417 tons of digestate/a. Considering the selling price of the digestate of 160 Yuan/t, there is a possibility to produce additional revenue of 0.87 million Yuan/a.

Table 1: Methane and fermented residue potential of the biomass produced in ZQY.

Substrate	Availability (t/a)	CH_4 potential (m^3 CH_4/a)	Fermented residue t/a
Municipal waste in ZQY			
Kitchen waste ZQY	3,269.15	189,415.55	2,970.23
Collectable agriculture residue in ZQY [13]			
Corn-cob-husks-silage	748.00	125,745.98	508.94
Flower residues (cash crop)	840.00	29,739.36	783.933
Wheat straw	1,630.00	241,622.07	1,154.37
Sum	6,487.15	586,522.96	5,417.48

3.3 Compost potential in ZQY

3.3.1 Evaluation of ZQY raw organic waste composition for composting

To evaluate the compost potential from ZQY the compostable waste fractions from kitchen waste, other organic waste, and fractions < 40 mm size were considered (see Fig. 1). The produced waste in ZQY had the typical waste characteristics of most developing countries, such as high moisture content and large organic fractions. This particular characterises of the waste has significantly high potential to hinder the incineration or direct heat recovery.

The composition of the total waste delivered shows that the fresh matter in the waste for both trials (spring and summer 2019) was strongly dominated by organics. The overall waste composition consists of three waste fractions with the potential to be composted and sums up to 63% of total generated waste (Kitchen waste, other organic waste, and < 40 mm fractions). Diversity in organic waste creates promising opportunities for optimal utilization of this waste, specifically concerning the production of high quality finished compost. Organic waste contains kitchen waste (nitrogen source) and garden waste (carbon source) guarantees the suitable C/N ratio needed for efficient decomposition. This is due to the fact that the quality of the compost produced depends largely on the level of the C/N start-up ratio and the quality of constituents within the mixture.

3.3.2 Evaluation of the chemical properties of ZQY raw organic waste

High levels of heavy metals represent an obvious concern when the compost is destined to be applied to food crops [14]. Similarly, the concentration of major elements such as nitrogen, phosphorus, and potassium in the produced organic waste plays a major role in deciding the quality of the compost. Heavy metals do not degrade during the composting process and always become more concentrated due to microbial degradation. Heavy metals in compost products are sourced from the raw materials that have been subjected to composting. In this context, analysis results showed that the resulting heavy metal and nutrients in ZQY raw organic materials were appropriate with the limits set by urban waste for agriculture use, China [15] and ordinance on the recovery of bio-waste on agricultural, forestry, and horticultural uses, Germany [16] (see Table 2).

Table 2: Heavy metal and nutrients concentrations of ZQY raw organic materials compared with Chinese [15] and German standards[16].

Parameter	Range in ZQY organic waste	Control standards for urban wastes for agriculture use in China	German standards BioAfV, 2017	
			Class A	Class B
Heavy metal concentration				
Pb mg/kg	2.94–26.70	<50	<150	<100
Cd mg/kg	< 0.40	<3	<1.5	<1.0
Cr mg/kg	6.48–69.10	-	<100	<70
Cu mg/kg	8.89–59.30	-	<100	<70
Ni mg/kg	2.39–26.40	-	<50	<35
Hg mg/kg	0.050–0.053	<2	<1.0	<0.7
Zn mg/kg	46.80–126.00	-	<400	<300
As mg/kg	1.00–2.60	<15	<10	<10
Nutrients concentration				
N%	1.79	>0.5	-	
P_2O_5%	1.13	>0.3	-	
K_2O%	1.33	>1	-	

The content of the heavy metals (Pb, Cd, Cr, Cu, Ni, Hg, Zn, and As) shown in Table 2 indicates that the concentrations of all seven heavy metals and the nutrient concentration were appropriate with the limits set by Chinese and German Standards. The low concentrations of heavy metals are because the waste was generated and collected from the rural area. Usually, the waste generated from rural areas has a very low percentage of heavy metals, in contrast to those that are generated in urban areas (high concentrations of heavy metals).

3.3.3 Economic feasibility of compost production in ZQY

To assess the economic benefits of composting in ZQY a model is proposed and the proposed model considers several facts and assumptions as following;

i. 38 tons/day of waste generation
ii. 100% waste collection efficiency
iii. 63% of the collected waste is compostable waste
iv. Composting duration of 12 weeks
v. 30% volume reduction in windrow piles during composting and
vi. The selling price of compost at 500 Yuan/ton.

With the above assumptions nearly 5825.4 tons/year of finished compost can be produced. Considering the finished compost has 50% of moisture content, nearly 2912.7 tons/year of finished dry compost can be produced from the waste generated in ZQY. This, in turn, can produce revenue of 1.46 Million Yuan/year from the sales of finished compost.

Table 3: Recyclables composition of municipal waste from ZQY city, an average of both trial analysis.

Recyclable fractions	Spring trial (%)	Summer trial (%)	Average (%)	Average dry matter (%)	Amount (ton/a)	Price/kg in Yuan [18]	Annual revenue (Yuan)
PPC	2.65	2.62	2.64	53.17	195	0.80	156,000.0
Glass	2.28	3.74	3.00	-	416	0.95	395,200.0
Plastic packaging	10.69	15.39	13.00	59.52	1063	0.70	744,100.0
Metal	1.85	1.01	1.43	-	198	0.90	178,200.0
Total revenue (price vary from city to city)							1,473,500.0

3.4 Recyclables potential in ZQY

3.4.1 Quantity of recyclables generated in ZQY

In the recyclable waste category, the composition of PPC is comparatively similar without any major seasonal variations (refer to Fig. 1). The glass composition is relatively higher in summer with 3.74% and 2.28% in spring. The plastic packaging fractions have dominating shares in the recyclable waste category with 10.69% in spring and 15.39% in summer. Nevertheless, the overall recyclable waste composition is lower compared to the average recyclable produced in the EU [12]. In the recyclable waste category, the composition of PPC is significantly lower not only in comparison with the different economic levels of the counties but also with the same economic level as well. PPC averages about 2.64% in ZQY in comparison with 9.5% in other middle-income countries [17], the source-segregation of PPC in ZQY explains it. However, the composition of plastic, glass, and metal almost similar in ZQY with other middle and low-income countries. The difference observed in the composition of ZQY can be related to (i) consumption pattern and geographical factors, (ii) waste management system, (iii) local regulations, (iv) methodology of waste characterization.

3.4.2 Economic feasibility of recycling the recyclables in ZQY

Due to the huge recycling potential as a resource, PPC, glass, plastic, and metals waste are subject to recycling obligations with specific quotas as legally stipulated by national guidelines. In the same light, China is currently planning to implement a source-separated collection system for recyclables waste. To this concern, this study helps to identify the value of those recyclable fractions and the economic feasibility of utilizing them. Table 3 shows the most notable valuable recyclable fractions that should be focused and their potential monitory benefits.

The recyclable fraction in Table 3 was sorted from the highly heterogeneous municipal waste from ZQY. Fraction like paper and plastic packaging due to their obvious surface and structural characteristic tends to attract moisture and dirt, with some portion of PPC reacting moisture up to 70%. Hence the separate waste collection is the primary prerequisite for successful waste recycling. However, in this feasibility analysis the recyclable waste produced is calculated according to their dry matter % enabling the most reliable cost estimations. Considering the waste generation rate of 38 tons/year and 100% collection efficiency over 365 days of the year, there is a potential to generate 195 tons/year of PPC, 416 tons/year of glass, 1,063 tons/year of plastic packaging, and 198 tons/year of metal. The amount of generated recyclables in ZQY has latent to produce total revenue of 1.47 million Yuan/year.

3.5 Total estimated waste value potential of the municipal waste produced in ZQY

Table 4 represents the summary of the total value potential of the municipal waste generated in ZQY. The obtained results depict that organic waste produced in ZQY has comparatively higher value potential than the rest of the available wastes. The total estimated value potential of municipal waste produced in ZQY is between 2.93–3.66 million Yuan/year (depending on the choice of treatment technology; Biogas + recycling or composting + recycling).

Table 4: Total estimated value potential of the municipal waste produced in ZQY.

Treatment method	Value product	Estimated value in million Yuan/year	Total revenue in million Yuan/year
Biogas	Heat and electricity	1.02	1.89
	Fermented residue	0.87	
Composting	Compost	1.46	1.46
Recycling	PPC	0.15	1.47
	Glass	0.39	
	Plastic packaging	0.74	
	Metal	0.18	

4 CONCLUSION

The study depicts the detailed value potential of the waste produced in ZQY town. The illustrated results have a high potential to serve as a basis for the municipal decision-makers and to use as a model study for analysing waste value potential in typical Chinese peri-urban settlement areas. From the environmental and economic viewpoint, composting and biogas production can be beneficial in treating the organic fractions of solid waste in ZQY. The overall, waste composition of ZQY and its chemical properties indicate that the composting and biogas production is an economical and sustainable alternative to treat the waste produced in ZQY. Similarly, recycling the recyclables also provides an opportunity to generate huge additional revenue. However the separate waste collection system is highly recommended to successfully recycle the recyclables. The self-reliance of waste management in the peri-urban region like ZQY can significantly reduce the overburden on bigger cities in their waste management. Additionally, the obtained results depict that organic waste produced in ZQY has a comparatively higher value potential than the rest of the available wastes. This gives major reasoning for making bio-waste management and value recovery a primary focus in ZQY. Conversion of organic waste in ZQY into compost and biogas can bring unparalleled economic, environmental, and social benefits and has a higher potential in developing sustainable waste management in ZQY by promoting a low-carbon circular economy.

ACKNOWLEDGEMENT

The authors acknowledge the Federal Ministry of Education and Research (BMBF) for financing this study via the PERIVAL – Waste to value chains in peri-urban environment project.

REFERENCES

[1] Dezhen Chen, L.Y.H.W.P.H., Pyrolysis technologies for municipal solid waste: A review. *Waste Management*, **34**(12), pp. 2466–2486, 2014.

[2] National Bureau of Statistics of China, *China Statistical Yearbook*, China Statistics Press, 2017.

[3] Lino, F.A.M. & Ismail, K.A.R., Analysis of the potential of municipal solid waste in Brazil. *Environmental Development*, **4**, pp. 105–113, 2012.

[4] Lino, F., Bizzo, W.A. & Silva, K.A.R.I.E.P., Energy impact waste recyclable in a Brazilian Metropolitan. *Resources, Conservation and Recycling*, **54**, pp. 916–922, 2010.

[5] Oazana, S. et al., A flexible control system designed for lab-scale simulations and optimization of composting processes. *Waste Management*, **72**, pp. 150–160, 2018.

[6] Duana, N., Lin, C., Wang, X.D., Zhang, X.J. & Hou, Y., Study on the effect of biogas project on the development of low-carbon circular economy: A case study of Beilangzhong eco-village. *Procedia Environmental Sciences*, **5**, pp. 160–166, 2011.

[7] Bilitewski, B., Wagner, J. & Reichenbach, J., *Best Practice Municipal Waste Management*, Umweltbundesamt: Dessau-Roßlau, 2018.

[8] Riber, C., Petersen, C. & Christensen, T.H., Chemical composition of the material fraction in Danish household waste. *Waste Management*, **29**(4), pp. 1251–1257, 2009.

[9] Towprayoon, S. et al., Waste generation, composition and management data. *Refinement to the 2006 IPCC Guidelines for National Greenhouse Gas Inventories*, 2019.

[10] Saeed, M.O., Hassan, N.M. & Mujeebu, M.A., Assessment of municipal solid waste generation and recyclable materials potential in Kuala Lumpur, Malaysia. *Waste Management*, **29**, pp. 2209–2213, 2009.

[11] Srivastava, V., Ismail, S.A., Singh, P. & Singh, R.P., Urban solid waste management in the developing world with emphasis on India: challenges and opportunities. *Reviews in Environmental Science and Bio/Technology*, **14**, pp. 317–337, 2015.

[12] Bidlingmaier, W., Sidaine, J.-M. & Papadimitriou, E., Separate collection and biological waste treatment in the European Community. *Reviews in Environmental Science & Bio/Technology*, **3**, pp. 307–320, 2004.

[13] Shunyi Bureau of Statistics, *The Statistical Yearbook of Beijing Shunyi 2018*, China Statistics Press: Beijing, 2018.

[14] Papadimitrou, E., Barton, J. & Stentiford, E., Sources and levels of potentially toxic elements in the biodegradable fraction of autoclaved non-segregated household waste and its compost/digestate. *Waste Management and Research*, **26**, pp. 419–430, 2008.

[15] Ministry of Environmental Protection of the People's Republic of China, Control Standards For Urban Wastes For Agricultural Use – GB8172-87, 1987.

[16] Bioabfallverordnung – BioAbfV, Verordnung über die Verwertung von Bioabfällen auf landwirtschaftlich, forstwirtschaftlich und gärtnerisch genutzten Böden, Bundesministeriums der Justiz, 2017.

[17] Edjabou, M.E. et al., Municipal solid waste composition: Sampling methodology, statistical analyses, and case study evaluation. *Waste Management*, **36**, pp. 12–23, 2014.

[18] Thomas Dorn, M.N.S.F., Circular Economy in China. *ISWA*, 2010.

SOLID WASTE MANAGEMENT AT AN INTERNATIONAL FULL-MARATHON RUNNING EVENT IN SOUTHERN THAILAND

SINEENART PUANGMANEE & MALIK SAEARLEE
Faculty of Science and Technology, Phuket Rajabhat University, Thailand

ABSTRACT

Marathon running is the most popular sporting event in the world. Therefore, marathon solid waste is an important aspect of these events. Many activities that support the runners produce much waste, which is difficult to control and manage. This paper aimed to study marathon solid waste management at an international full-marathon running event, along with the support activities. The data included the types and volumes of waste in addition to waste storage, collection, transfer, and disposal. The sample site was an international full-marathon event in southern Thailand. Two groups of activities were identified. The first group included 11 activities at service points directly concerned with the runners. The second group included two activities not related with marathon running, i.e. an expo zone and food zone. All waste was classified into general waste, compostable waste, recyclable waste, and hazardous waste for each group activity. The total amount of waste from all activities was 2,702.70 kg, that included 2,527.20 kg and 175.50 kg that were related and unrelated to marathon running, respectively. Moreover, it was noted that unsuitable protective equipment was worn by the workers. Some containers and vehicles did not follow the guidelines according to the laws; however, the government allowed them to work. A waste recycling facility received 458.10 kg of recyclable waste. General waste, compostable waste, and hazardous waste (2,244.60 kg) were transferred without separation for disposal by incineration. Although the organizers tried to control and manage the waste at the event, some of the waste handling procedures were incorrect and ineffective. In conclusion, the organizers and participants need greater awareness of eco-friendliness for an eco-marathon.

Keywords: international full-marathon running event, marathon running, solid waste management.

1 INTRODUCTION

Recently, most people exercise frequently and play sports. One sport that is the most popular is marathon running. Many marathon running events at the local, national, and international level are conducted worldwide. Some running events are world famous, such as the London Marathon, BMW-Berlin Marathon, Boston Marathon, Chicago Marathon, New York City Marathon, and Tokyo Marathon [1]. It is a challenge, especially for runners who want to beat the previous records. Normally, a running event includes the organizers and the owners of the running area who organize and manage the event to support the runners and spectators. All of these events produce waste as a result of the event. During the event, waste is scattered in the area of the event and along the running route which can be an ugly sight. In addition, waste from food, packages, and goodie bags get left behind by the spectators and runners [2]. In the Boston Marathon of 2017, 1.4 million paper cups were distributed to the runners and 171,380 paper brochures and visitor's guides were printed [3]. In the London Marathon of 2018, the Westminster City Council collected 5,200 kg of rubbish and 3,500 kg of materials for recycling that included 47,000 plastic bottles from the street [2]. Moreover, some international marathons do not control or manage the waste correctly even though they are aware of the environmental impact from the event.

Thailand has many marathons at both the local and national levels that produce a lot of waste. Therefore, we aimed to study waste management at an international full-marathon running event in southern Thailand and collect data on the support activities, running matches, types and quantities of waste, the storage, collection, and transfer/transport of the waste, and the disposal of solid waste materials.

2 METHODOLOGY

This survey research focused on solid waste management at an international full-marathon running event (IFmRE) that was set in southern Thailand. IFmRE replaces the true name of the marathon running event to protect the identity. This event is the biggest marathon in South-East Asia, and meets the highest international standard that is also the qualifier for the Boston Marathon. The event is professionally scheduled by Sportstats Asia and it is certified by the Association of International Marathons and Distance Races.

2.1 Area site

The IFmRE was set in a district of Phuket Island in southern Thailand. The number of runners exceeded 13,000 participants from more than 50 countries and took place during 7–9 June 2019.

2.2 Survey method

2.2.1 Surveys and interviews

Surveys and interviews were conducted with the organizer, the owner of the area for the marathon, and employees to determine the overall format of the running event, the related and unrelated activities of the running events, and the waste management of the IFmRE. This study documents the types and quantities of waste and the storage and collection of waste in terms of containers, employees, and equipment. Also studied were the transfer/transport and disposal in terms of employees, periods of time for transfer/transport, equipment, vehicles used, and the disposal site.

2.2.2 Sampling frames

The types and quantities of the IFmRE waste (IFmREW) were studied by sort segregation and the waste was weighed from all activities of the marathon that covered three running matches that consisted of a half-marathon (21.0975 km), marathon (42.195 km), and marathon relay (42.195 km). On 9 June 2019, the number of runners was around 7,304 people. The waste was collected from 04:00 a.m. to 12:00 p.m. The IFmREW that was collected included waste related directly to the running event and waste that was unrelated to the running event.

3 RESULTS AND DISCUSSION

The results of this article describe the activities and running matches, types and quantities of waste, waste storage and collection, and waste transfer/transport and disposal at the IFmRE.

3.1 Marathon running event activities

A total of 13 activities produced waste materials. Eleven activities produced waste that was directly related to the running event (IRFmRE) and two activities produced waste that was unrelated to the event (IUFmRE) (Table 1).

Table 1: Activities and sources of waste.

No.	Activities and sources of waste	
1	International related full-marathon running event (IRFmRE)	1.1 Souvenir service point
		1.2 Water service point
		1.3 Relaxing massage service point
		1.4 Food service point for runner
		1.5 Start/stop point
		1.6 Sponsor service point
		1.7 Winner register point
		1.8 First aid point
		1.9 Luggage drop-off service point
		1.10 Water service point along the running route (running route was used in 3 running matches)
		1.11 Toilet service point
2	International unrelated full-marathon running event (IUFmRE)	2.1 Expo zone
		2.2 Food zone

3.2 Types and quantities of solid waste

The total amount of waste was 2,702.70 kg which consisted of 2,527.20 kg of IRFmRE waste (IRFmREW) and 175.50 kg of IUFmRE waste (IUFmREW) (Fig. 1). The rate of waste based on the amount per person from all activities was 0.37 kg/person/day. The types and quantities of the waste were also categorized into four types: general waste, compostable waste, recyclable waste, and hazardous waste (Figs 2–4).

The amount of general IRFmREW was the greatest which consisted of paper (wax cups, bowls, and plates), sponges, and plastic bags. The results of paper cups (wax cups) were similar to reports by Mahmud [4] and Mervosh [5]. The compostable waste included coconuts, food waste, and fruit peelings from bananas and watermelons. The results of fruit waste materials were similar to a report by Mahmud [4] on the annual Singapore marathon and a report on the shrinking carbon footprint event at the Salzburg Marathon [6] where banana peels were found. In the category of recyclable waste, the amount of single-use water bottles was similar to the reports by Mahmud [4], Smithers [7], United Nations Environment Programme [8], Mervosh [5], Lewis [9], and Cheung [2]. The amount of single-use plastic cups was similar to the reports by the United Nations Environment Programme [8] and Lewis [9]. The amount of cardboard materials was similar to a report from the Monterey Bay Half Marathon [10]. Finally, the hazardous waste category found infectious waste materials, such as toilet rolls, Accu-Chek FastClix and absorbent cotton/paper contaminated with blood, masks, and medical plaster bandages. Toxic waste materials included containers of pain relief medicine for relief of muscular aches and pain in the forms of spray, cream, and gel, and bottles for 70% alcohol.

In the IUFmREW, general waste was the highest amount which included plastic bags, paper materials including wax cups, bowls, and plates, wet paper and wet cardboard, foam boxes, plastic spoons and boxes, and packs of energy gel. These materials were similar to the findings by Mahmud [4]. In the category of compostable waste, food materials were found and in the category of recyclable waste, single-use plastic bottles were found. These results were similar to those reported by Mahmud [4], Smithers [7], Mervosh [5], Lewis [9], and Cheung [2]. The amount of aluminum cans was similar to the report by Mahmud [4] and the amounts of glass bottles and single-use plastic cups were similar to the reports by the United Nations Environment Programme [8] and Lewis [9]. The amount of cardboard materials was similar to the Monterey Bay Half Marathon report [10]. The hazardous waste materials included spray bottles, batteries, and print cartridges (Fig. 2). The amounts of waste materials are illustrated in Figs 3 and 4.

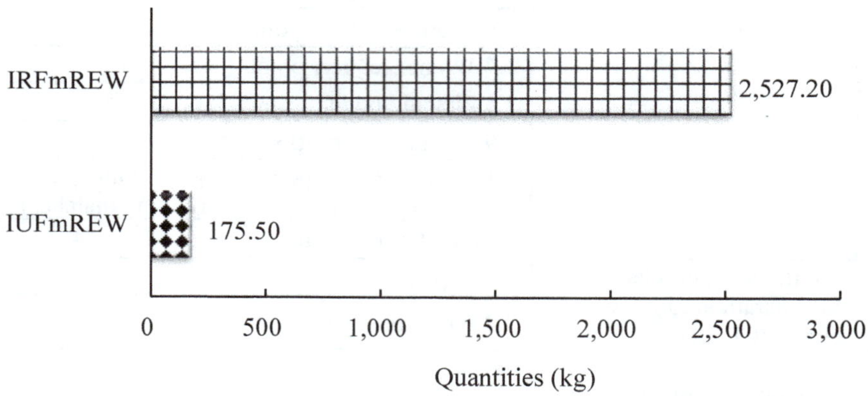

Figure 1: Types and quantities of IFmREW.

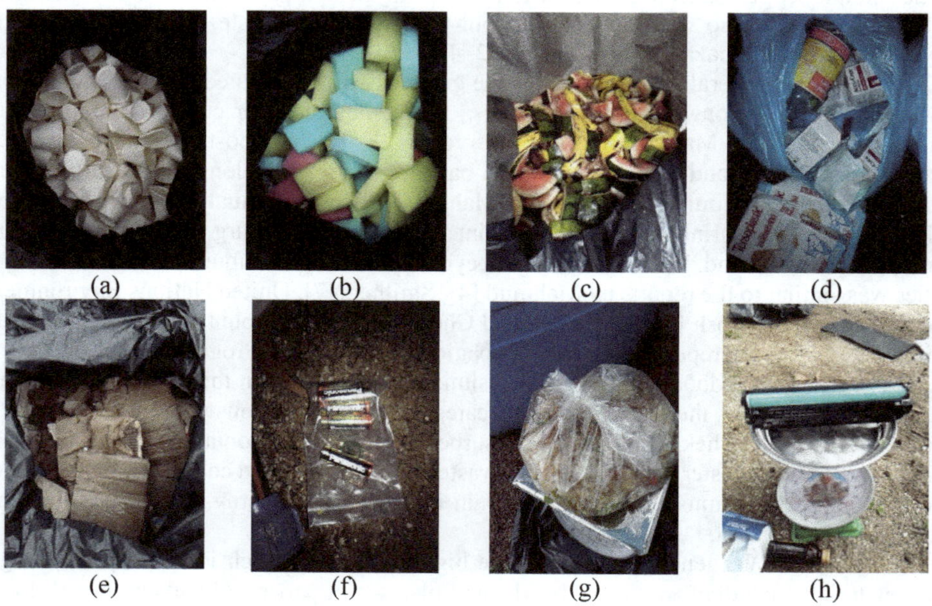

Figure 2: (a)–(d) Waste materials from IRFmREW; and (e)–(h) Waste from IUFmREW.

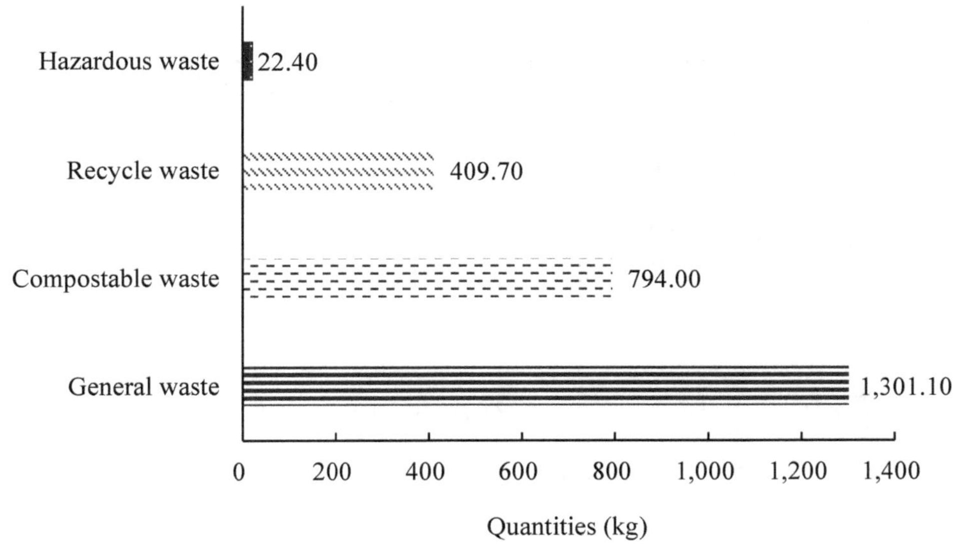

Figure 3: Types and quantities of IRFmREW.

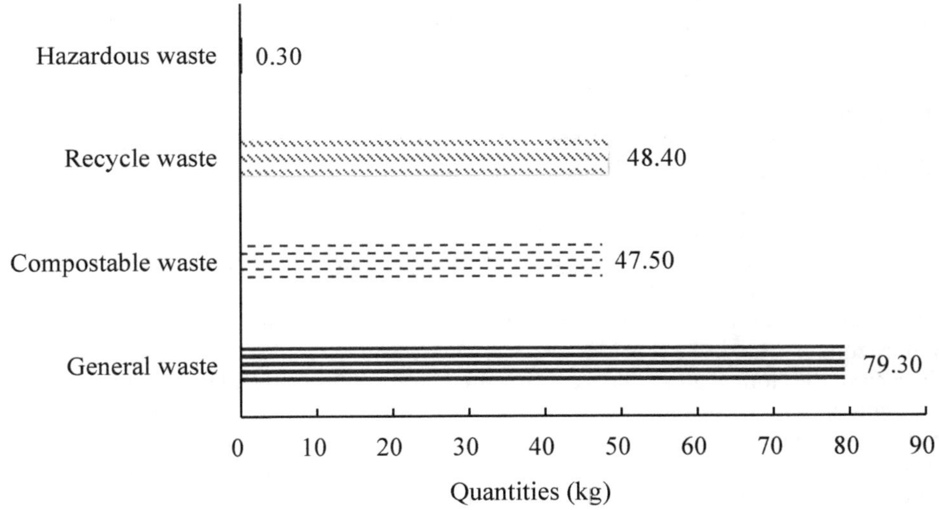

Figure 4: Types and quantities of IUFmREW.

Some waste at the marathon event was found to be separated at the sources for disposal at the end of pipe. These waste materials were separated into two types that consisted of general waste (2,244.60 kg), such as wet plastic materials (bags, boxes, bowls, and plates), wet paper materials (boxes, wax cups, plates, and bowls), foam (boxes, bowls, and plates), and recyclable waste (458.10 kg) that included clear plastic PET bottles, colored plastic bottles, and paper and cardboard materials (Fig. 5).

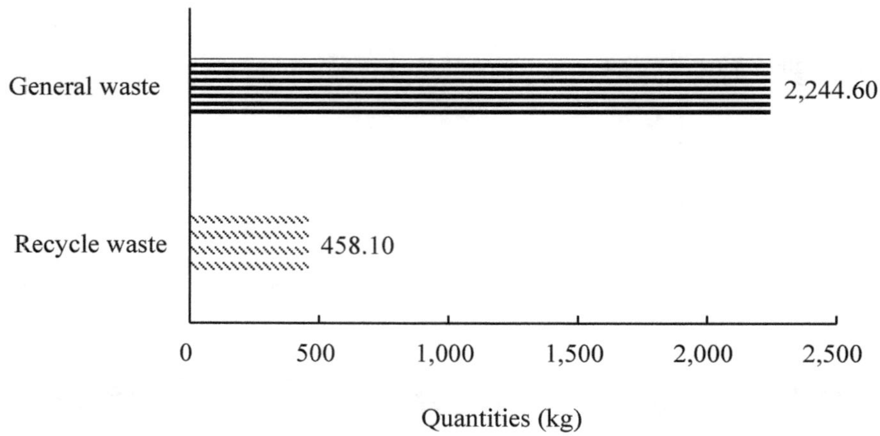

Figure 5: Types and quantities of IFmREW disposal.

3.3 Waste storage and collection

Most waste materials were not separated at the source. The various waste materials were deposited together into black plastic bags that were placed inside steel frames, plastic buckets, or baskets (Fig. 6). Waste materials at some points were separated by waste workers at the sources into the two types of general waste and recyclable waste for easier disposal at the end. All waste materials from the running event were moved to temporary waste storage areas: Storage 1 and Storage 2.

At Storage 1, the waste was collected from souvenir service points, food service points for the runners, winner registration point, first aid points, luggage drop-off service points, toilet service points, and waste from the running route. At Storage 2, waste was collected from the water service points, relaxing and massage service points, start/stop points, sponsor service points, and food zones. Storage 1 and Storage 2 areas were unsuitable according to the laws because they were not buildings and did not have roofs for protection from rain in the event of rainfall. The two storage areas were simple stalls in an area near the activities zone.

In each of the storage areas, the waste materials were combined and classified by workers for separation into two kinds of waste: general waste destined for an incinerator and recyclable waste destined for a recycling shop.

The waste workers who collected the waste can be classified into three groups: 12 employees employed from an outsourced company; five employees (garden department workers) of the owner of the marathon event area; and three temporary workers that the owner of the marathon running event area employed for waste collection. Seventeen workers kept the waste from all activities within the running event area, except the running route. The three temporary workers collected the waste and moved it from the running route by local vehicle. They then moved all of the waste materials to the two temporary waste storage points and waited for transfer of the waste to a disposal site around two days later.

While working, they wore unsuitable personal protective equipment. They did not wear health masks, gloves, rubber boots, plastic apron, medical caps, or eye goggles (Fig. 7). These practices were in conflict with the Ministerial Regulation on General Waste Management B.E. 2560 (2017) [11].

(a) (b) (c)

Figure 6: Containers of waste. (a) Steel frame for hanging black plastic garbage bag; (b) Plastic bucket with black plastic bag; and (c) Basket with black plastic bag.

Figure 7: Personal protection equipment suits of the waste workers.

3.4 Waste transfer/transport and disposal

The waste materials at the two storage points remained there for around two days. The general waste and recyclable waste were then transported for disposal by six employees from a private company sanctioned by the government. They moved the waste materials using two pickup trucks following the completion of the running event. The local vehicles and pickup trucks were in conflict with the guideline (Fig. 8) [11] because the vehicles were unsuitable for hauling waste materials because they could leak, spill, or allow the waste to fly away during transfer and transport which would contaminate the environment. According to law, the vehicles must be made from a durable material and structure that is easily cleanable and can be closed to prevent waste from leaking and flying away from the transport vehicle. At the end of the process, the waste materials were transported for disposal at an incinerator or to a recycling center.

(a) (b)

Figure 8: Waste transfer and transport vehicles. (a) Local vehicles move waste from running route; and (b) The pickup truck moves waste to disposal.

The general waste materials were moved to an incinerator plant and the recyclable waste was transported to a recycling shop that met the requirements of the Ministerial Regulation on General Waste Management B.E. 2560 (2017) [11]. The disposal recycling methods were similar to reports from the AACR Philadelphia Marathon 2017 [12], Swiss City Marathon 2018 [13], Monterey Bay Half Marathon 2017 [10], BMW Berlin Marathon 2019 [14], Sanlam Cape Town Marathon 2019 [13], and the London Marathon 2019 [15]. The Philadelphia Marathon disposed waste by recycling, donation, and landfill, the Swiss City Marathon disposed waste by recycling, the Monterey Bay Half Marathon disposed waste by recycling, composting, donating, and landfill, the BMW Berlin Marathon put the waste into recycling processes using normal recycling and thermal recycling (energy-from-waste), and donating, the Sanlam Cape Town Marathon disposed waste by recycling or composting, and the London Marathon used recycling for disposal of the waste.

The waste disposal flow chart of the IFmREW is shown in Fig. 9. Overall, the waste materials were not separated. In fact, these waste materials could have been separated at all points for easy classification of the waste for disposal. The organizer should put specifically labeled and suitable containers for depositing the waste according to the type of waste for easy handling before transfer to the temporary waste Storage 1 and Storage 2. Some waste at some points in the storage areas were separated into recyclable waste and general waste and was mixed with compostable and hazardous materials. The waste materials were transferred and transported for disposal at a recycling shop or incinerator using pickup trucks. The methods used for waste disposal followed the laws but the vehicles were in conflict with the laws. In addition, all waste materials were produced by the main organizer and sponsors. If they are aware of the environmental impact, they must control and manage the waste at each activity at the sources with the concept of reduce, reuse, and recycle.

We have these suggestions.

1. Change and reduce some types of containers for food and drinks. All single-use plastic and foam items, such as plastic spoons, plastic boxes, wax cups, plastic bottles, and plastic cups should be replaced with compostable or biodegradable cups made from maize straw. Plastic bottles should be made from 100% or 50% recycled plastic and reduce the unrecyclable plastic packaging for food and drinks.

2. Change and reduce some types of fruits since some fruits are difficult to destroy and are not compostable, e.g., coconuts. Coconuts should not be available or the supply of coconuts should be greatly reduced.
3. Add more types of containers for the depositing of waste and use at least 4 types, i.e. general waste, compostable waste, recyclable waste, and hazardous waste for separation of the waste at the source site.
4. Separate the waste for suitable disposal, i.e. recycling, compost, donation, landfill, and incineration.
5. The temporary waste storage facility should be a durable structure with a roof to protect the waste from rainwater. When waste materials become wet, they can produce a leachate that contaminates the soil, reservoirs, and streets during waste transfer/transport. Also, some wet recyclable waste materials cannot be moved to a recycling center.
6. The waste handlers need to wear suitable personal protective equipment according to the guideline.

Finally, the vehicles for waste transfer/transport should be suitable and meet the requirements according to the law.

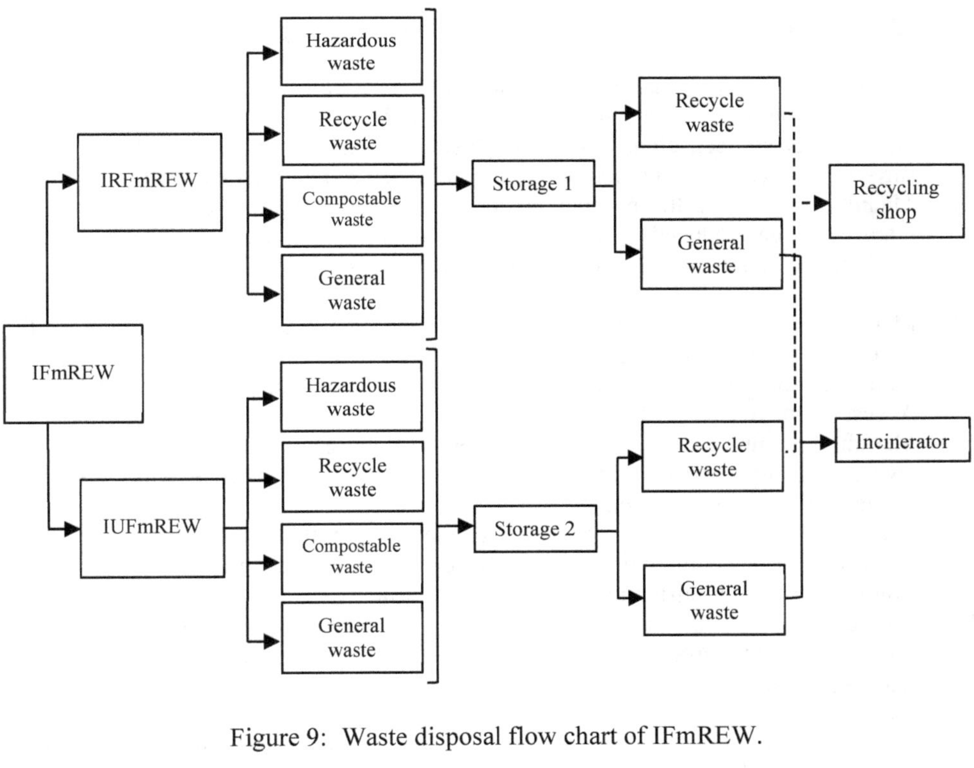

Figure 9: Waste disposal flow chart of IFmREW.

4 CONCLUSIONS

The IFmREW were separated into two types of waste. (1) IRFmREW included hazardous waste, recyclable waste, compostable waste, and general waste. The main sources of waste were from 11 activities (souvenir service points, water service points, relaxing massage service points, food service points for the runners, start/stop points, sponsor service points, winner register point, first aid points, luggage drop-off service point, water service points along the running route, and toilet service points. (2) IUFmREW included hazardous waste, recycle waste, compostable waste, and general waste. The main sources of waste were from the expo and food zones. The proportion of IRFmREW was much greater than IUFmREW. The waste storage points and collection were improper. The personal protective equipment of the waste workers was unsuitable and vehicles used for the transfer and transport of the waste materials were in conflict with the guideline. The practices of using an incineration plant for general waste and recycling shops for recyclable waste were lawful according to the laws of Thailand. Even though an international full-marathon running event is conducted over a short period of time, it can produce much waste. The organizer, sponsors, owner of the running event area, runners, and participants need to be aware of the impact on the environment and use positive measures to protect the ecosystem.

ACKNOWLEDGEMENT

The authors would like to thank Phuket Rajabhat University for their financial support for the registration and carfare.

REFERENCES

[1] Abbott World Marathon Majors. www.worldmarathonmajors.com. Accessed on: 12 Feb. 2020.

[2] Cheung, H., London Marathon: How do you reduce the environmental impact? 2019. www.bbc.com/news/science-environment-48064102. Accessed on: 10 Feb. 2020.

[3] How to stay eco-conscious during the Boston marathon. https://tataandhoward.com/stay-eco-conscioys-during-boston-marathon/. Accessed on: 12 Feb 2020.

[4] Mahmud, A.Z., See the number of cups thrown: Can the waste produced during marathons be reduced? 2018. www.channelnewsasia.com/news/singapore/wste/produced-marathons-runs-races-organisers-go-green-10257948. Accessed on: 19 Nov. 2019.

[5] Mervosh, S., Seaweed pods, anyone? Marathons get creative to stop littering the streets, 2019. www.nytimes.com/2019/04/30/sports/marathons-plastic-water-bottles.html?utm_source=headtopics&utm_medium=new&utm_campaign=2019-04-30. Accessed on: 10 Jan. 2020.

[6] Marathon running's shrinking footprint event, Salzburg marathon, 2019. www.continental-tires.com/car/about-us/media-services/gripworld/events/2019-05-20-fussabdruck. Accessed on: 10 Feb. 2020.

[7] Smithers, R., Single use plastics bottles banned from half marathon in London, 2018. www.theguardian.com/environment/2018/sep/12/single-use-plastic-bottles-banned-from-half-marathon-in-london. Accessed on: 10 Jan. 2020.

[8] United Nations Environment Programme, Xiamen marathon runs towards plastic-free future with #CleanSeas, 2018. www.unenvironment.org/news-and-stories/press-release/xiamen-marathon-runs-towards-plastic-free-future-cleanseas. Accessed on: 19 Jun. 2019.

[9] Lewis, S., Thousands of seaweed pods will replace single-use plastic at the London Marathon, 2019. www.cbsnews.com/news/london-marathon-thousands-of-seaweed-pods-willreplace-single-use-plastics/. Accessed on: 10 Dec. 2019.

[10] Montery Bay Half Marathon, Zero-waste-event, 2017. www.montereybayhalfmarathon.org/zero-waste-event/. Accessed on: 3 Feb. 2020.

[11] Ministerial Regulation, General waste management B.E 2560. http://laws.anamai.moph.go.th/ewtadmin/ewt/laws/main.php?filename=2RLaws. Accessed on: 10 Dec. 2019.
[12] AACR Philadelphia marathon, Green Initiatives, 2017. www.philadelphiamarathon.com/info/green-initiatives. Accessed on: 8 Feb. 2020.
[13] Marathon running's shrinking footprint event, Swiss city marathon and Sanlam Cape Town Marathon, 2019. www.continaental-tires.com/car/about-us/media-services/gripworld/events/2019-5-20-fussabdruck. Accessed on: 10 Nov. 2019.
[14] Marathon running's shrinking footprint event, Swiss City marathon and Sanlam Cape Town Marathon, 2019. www.continaental-tires.com/car/about-us/media-services/gripworld/events/2019-5-20-fussabdruck. Accessed on: 10 Nov. 2019.
[15] Kottasova, I., Seaweed pouches will replace thousands of plastic bottles at the London marathon, 2019. https://edition.cnn.com/2019/04/26/business/london-marathon-seaweed-water-bottles/index.html. Accessed on: 8 Feb. 2020.

IMPROVING THE ATTITUDE AND REACTION TOWARDS MUNICIPAL SOLID WASTE MANAGEMENT IN MOZAMBIQUE

AMAD H. A. GANI[1], OLAIDE M. ADEROJU[2], ANTÓNIO G. DIAS[2] & ANTÓNIO A. R. MONJANE[1]
[1]Universidade Pedagógica de Maputo (UPM), Mozambique
[2]Department of Geosciences, Environment and Spatial Planning, Faculty of Science, University of Porto, Portugal

ABSTRACT

The understanding of a sustainable municipal solid waste (MSW) management plan requires proper education and awareness by all the stakeholders. Uncoordinated disposal pattern of MSW, nonchalant attitude of the authorities and stakeholders, weak legislation and policies among others have contributed to the increasing human health risk, greenhouse gas emissions, as well as the degradation of the environment at large. Currently, Mozambique is faced with the problems of indiscriminate dumpsites in most of its major cities. Hence, this study aims at enlightening the stakeholders on their roles towards a sustainable MSW management through an education and outreach model. The methodology employed in this study was an integrated concept with four main components, which includes principal actors, routes, procedures, and people. However, in conclusion, the integration of this conceptual model into the existing MSW management framework in Mozambique is likely to promote environmental consciousness amongst the people. Furthermore, aesthetic, pollution-free, as well as a sustainable environment is assured.

Keywords: education, environment, management, MSW, outreach, sustainability.

1 INTRODUCTION

The problem of municipal solid waste management (MSWM) has emerged as one of the trending environmental issues in the world, especially for developing countries. Aderoju et al. [1], stated that the world will continue to generate municipal solid waste (MSW) in high quantity and varieties as a result of rapid urbanization, industrial development and socio-economic impact, which in turn has become an environmental issue that requires urgent attention by every nation. In a study by Nnaji and Utsev [2], it was reported that the overstretching of waste management facilities and the inability of waste management authorities, to cope with the volume of solid waste generated is associated with increasing urbanization in major cities. It is unfortunate that the standards of waste management are still poor and outdated in many developing countries, with poor documentation of waste generation rates and its composition, inefficient storage and collection systems, disposal of municipal wastes with toxic and hazardous waste, indiscriminate disposal or dumping of wastes and inefficient utilization of disposal site space [3].

The issues of creating awareness and proper education on MSW management and disposal, ethics plays a good role in attaining environmental sustainability. Aderoju et al. [4] suggested that MSW sensitization and environmental education programs should be carried out in both urban and the rural communities. Moreover, Jackson [5] reported that many people are unable to exercise deliberate choice because they find themselves locked into unsustainable patterns caused by habits, routines, lack of knowledge, institutional structures, inequalities in access, social expectations and cultural values. In many African communities, lack of awareness and proper education on integrated waste management has contributed to the pattern of disposal of MSW and poor management practices. Several studies have reported that many African cities are still deficient in the proper management of their solid waste. For instance, many Mozambican cities are no exception to this whereby most of their

WIT Transactions on Ecology and the Environment, Vol 247, © 2020 WIT Press
www.witpress.com, ISSN 1743-3541 (on-line)
doi:10.2495/WM200051

MSW are being buried, burnt, or disposed haphazardly in the environment. According to INE (Instituto Nacional de Estatistica de Moçambique) [6], 46% of the Mozambican population engage in the combustion of MSW within the perimeter of their homes, 23% bury, while 19% dispose of their MSW on open land. Concerns were raised in the past by the WHO [7] on poor sanitation and unsatisfactory waste management in African societies in a way to mitigate human health risk. According to the IPCC report of the 12th season of the intergovernmental panel on climate change [8], methane emission from landfill was estimated to account for 3% to 19% of the anthropogenic sources in the world. Moreover, UNWSSD [9] reported that the theme of the World Summit on Sustainable Development in Johannesburg, South Africa was on waste minimization, recycle, reuse and safe disposal of waste to minimize pollution of the environment.

Overcoming public attitudes and unsustainable behaviour requires effective communication, a broad public understanding of the requirements of solid waste management and active participation of all relevant stakeholders throughout all project stages [10]. One of the merits of active participation was reported by Rousta and Dahlén [11], that inhabitants' participation in waste separation schemes seems crucial for the improvements needed to achieve a better ambitious material recycling goal. According to Sallwey et al. [12], it was reported that illiteracy and low education levels pose a challenge for waste management services and awareness campaigns. However, the initiatives taken by the international stakeholders are often ineffective due to poor coordination and thus have not fostered much sustainable changes in waste management [13]. The transformation of the society to gain environmental and sanitary awareness is dependent on the level of education and outreach. The attitude of the Mozambican people in the major cities on waste disposal pattern and sanitation ethics has become an overburden on the authorities and the communities at large as it adverse effects, as reported by countless researchers, threaten human health risk. In this view, the study aims at enlightening the stakeholders on their roles towards a sustainable MSW management through an education and outreach model. This study is likely to bridge the gap of awareness on environmental ethics and solid waste management in the major cities in Mozambique. Furthermore, the study is likely to improve the mentality, behavior and practice of the inhabitants of the communities towards attaining a sustainable and aesthetic environment.

2 STUDY AREA

The Republic of Mozambique is located on the latitudes 10° 27'S and 26° 52'S and longitudes 30° 12'E and 40° 51'E, with a surface area of 799,380 km^2 [14]. Also, it is bordered on the north by Tanzania, on the west by Malawi, Zambia, Zimbabwe, and South Africa, and on the south by Swaziland and South Africa, with its territory bordered by the Indian Ocean, in a span of about 2,515 km to the east. Mozambique has a population density (see Fig. 1), that is predominantly a rural, with an estimated population at 27,909,798 inhabitants [6]. However, according to Muchangos [14], a little above 23% of the entire population resides in urban areas. The economy of Mozambique is dependent on agriculture, in which subsistence type of farming is the major practiced.

The geographical situation and history of this country marked by various migratory processes result in a heterogeneous population group, with multicultural and multiethnic characteristics. Furthermore, for economic reasons, Mozambique is a member of the Southern African Development Community (SADC) group of countries.

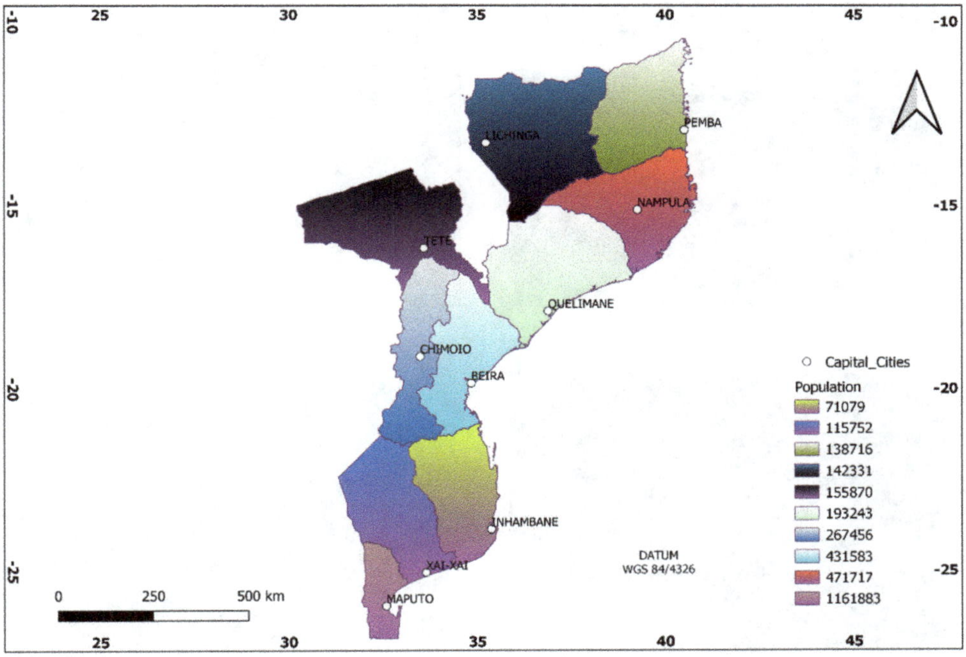

Figure 1: Population density map of Mozambique. *(Source: Adapted from [6].)*

3 STATUS OF MSW MANAGEMENT IN MOZAMBIQUE

The ideal practice in Mozambique on MSW management is for MSW to be collected, transported, and deposited in the municipal open containers by a public–private partnership, managed entirely by municipal city agencies, municipal town's agencies and govern of district administrations.

The national collection coverage ranges between 40–65% [15]. However, according to Tas and Belon [16], it was reported that Maputo's collection coverage increased to about 82% as compared to other major cities. Hence, the collection of waste from the local dumpsites in the inner city of Maputo increased profoundly from 76 million kilograms in 2007 to 253 million kilograms in 2014 [12]. However, there are setbacks affecting this effective and efficient plan includes, soft implementation of policies, low coverage of MSW collection, low financial support, absence of modern technological equipment, and ultimately, low awareness and education on waste management initiative. The reality of the environmental issues caused as a result of poor MSW management program in Mozambique is as shown in Fig. 2. INE [6] reported that at the national level, the rate of collection of household waste by municipal or administrative authorities is 7.6%, while 1.1% is collected by private companies.

Also, MITADER (Ministério da Terra, Ambiente e Desenvolvimento Rural) [17] showed that Maputo city produces the most MSW, with 415,005 tons per year of the country's total production and Inhambane city produces the least, with 17,520 tons per year (see Fig. 3). Individual municipalities, and private companies under a contract with the authorities also collects MSW from markets, restaurants, and other larger clusters in the community.

Figure 2: MSW reality in Mozambique. (a) Drainage dumping; (b) Street littering; (c) Illegal dump site; and (d) Indiscriminate dumping. *(Source: Authors' field investigation.)*

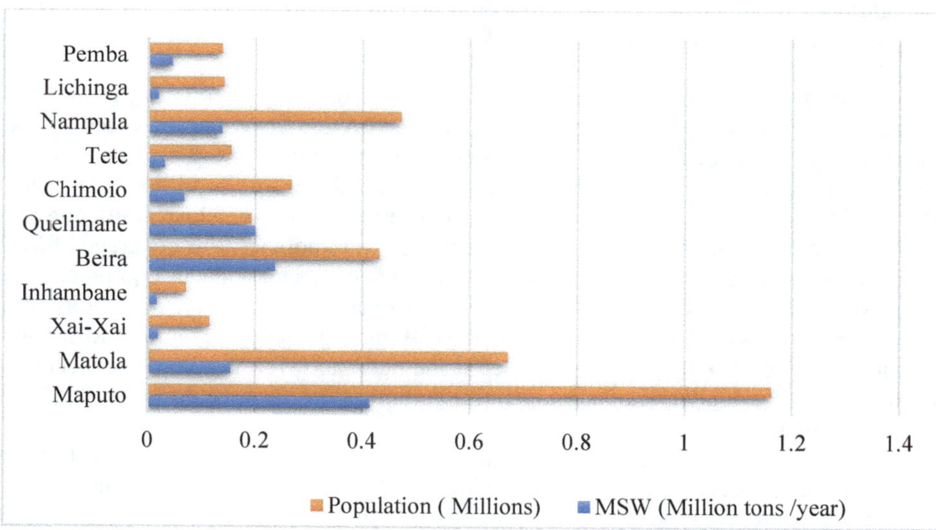

Figure 3: MSW generation in the capital cities of Mozambique. *(Source: Adapted from [17].)*

However, private companies, and few NGOs such as AMOR (Associação Moçambicana de Reciclagem) among few engage in waste segregation into waste categories like paper/cardboard, glass, plastic, metals, and others, for profit purposes. Cart pushers collect waste in places where there is no car access due to the absence of roads. In addition, the informal sector of the MSW management is clearly involved in the picking valuable and reusable materials from dumpsites for the purpose of recycling and useable. This is usually carried out by waste pickers otherwise known as "catadores", who earn their livelihood by collecting valuable materials from dumpsites and other places and selling it to vendors in the recycling industry.

The integration of waste pickers into the MSW management cycle is very significant as their contribution is likely to improve the recycling of MSW recycle in the many cities in other hand to clean the cities, helping the government. Again, it was reported in AMOR [18], and Buque and Ribeiro [19] that in a way to support informal collection, AMOR installed Eco-Points throughout Maputo to purchase recycled goods from private persons and waste pickers and resell to the recycling sector. Sallwey et al. [12] also reported that AMOR estimated that approximately 350 waste pickers sold about 300,000 kg of waste per month to these collection points in 2011.

4 EXISTING MSW MANAGEMENT LEGISLATIONS IN MOZAMBIQUE

The existing legislation and policy framework for solid waste management in Mozambique is mainly concerned with the requirements for licensing and auditing of public and private institutions involved in solid waste management [12]. The management of MSW remains the responsibility of the municipals, townships or district administrations. However, in the rural communities, particularly, MSW is usually managed by individuals, which is characterized by open dumping, burying and combustion which has a negative impact on human health and the environment at large. This prompts the government of Mozambique, to issue Decree No. 94/2014, in all the major cities on MSW management [20]. Although there is an existing Master Plan for the MSW in Maputo, the capital city of Mozambique, which is "Resolution No. 86/AM/2008 of 22 May 2006 [21], MSW Cleaning Posture in Maputo Municipality". This document defines the Maputo City cleaning system, integrating the MSW screening and management of the components. It applies to all public and private activities that directly or indirectly influence the cleanliness component of the municipality. It aims to establish the framework of general principles and standards of the MSW cleaning system.

4.1 Decree No. 94/2014, of 31 December, Regulation on Municipal Waste Management

The Decree No. 94/2014 [20] was established as a guideline to address one of the most serious urban environmental problems in Mozambique. This decree at the national level sets forth its principles, objectives, instruments, as well as guidelines regarding integrated solid waste management which include hazardous waste, the responsibilities of people and the government. The two most significant sections of this Decree No. 94/2014 [20] are Articles 4 and 8, respectively. Article 4, states that "all public and / private entities engaging in MSWM should develop and implement an integrated solid waste management plan that is managed based on the principles of the waste management hierarchy". Article 8 further states that it is imperative to promote good municipal solid waste management practices (recycling, composting, selective collection and landfill), in coordination with other public institutions, civil society organizations and the private sector. It is the obligation of public or private entities that carry out activities related to waste management to draw up a waste management plan they manage before the start of their activity. The mandate of these public and / private entities includes solid waste plans; semi-annual solid waste inventories and reporting systems; environmental monitoring and inspection; environmental education and the national register of hazardous waste operators.

5 STRUCTURE OF MSW MANAGEMENT IN MAPUTO CITY

The management of MSW in Maputo city utilizes a specified framework (see Fig. 4), whereby the Maputo City Municipal Council (MCMC) is responsible for sweeping and cleaning the municipal streets and the collection of all MSW produced in the city. The per capita generation of MSW per individual was estimated at 1 kg per person daily in the inner city in Maputo compared to 0.56 kg per person daily in the suburban areas [22].

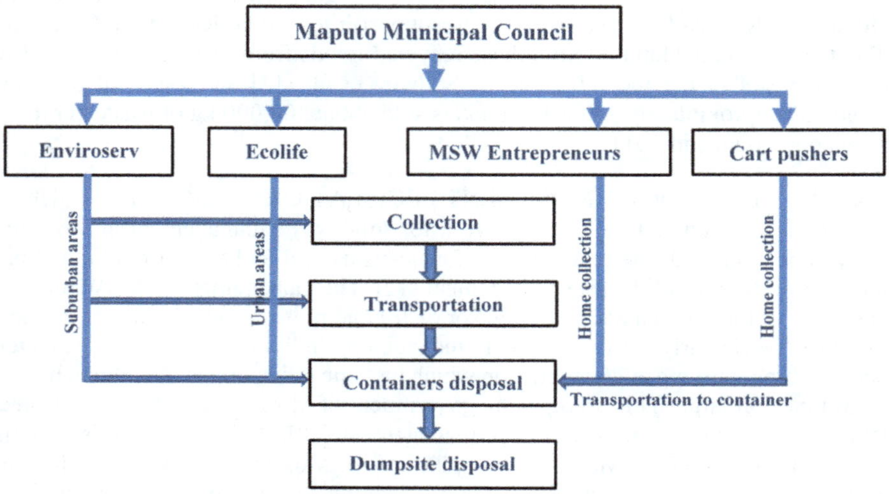

Figure 4: MSWM structure in Maputo city.

The MCMC through public private partnership (PPP) engaged two companies which include: the Ecolife, which collects wastes in the urban zones, and Enviroserv, which collects wastes in the suburban zones, particularly in markets, and transport terminals [23]. These two companies utilize 1.1 cubic meters of plastic containers with lids (Ecolife), and 6 to 10 cubic meters of metal containers (Enviroserv) and transports the collected MSW for safe disposal at the municipal Hulene dumpsite. It is unfortunate that the collection services in Maputo varies according to the level of income based on the income level and status of the neighbourhood. The MSW entrepreneurs also known as private micro-enterprises, licensed to collect household wastes, representing public–private partnerships (PPP), are licensed to collect at restaurants and other high-waste production sites. These private micro-enterprises also collect according to requests from any individual or institution that has enough MSW to dispose of if they pay the company for transportation costs. Also, the private micro-enterprises do not have the jurisdiction to transport the MSW collected to the municipal dumpsite. Instead, the MSW collected is transported to a private dumpsite to separation into different components for the sale of valuable materials. In addition, there exist cart pushers who are individual waste collectors in suburban areas that are not within the reach of the companies collecting for the MCMC. The cart pushers are also regarded as a member of the informal sector in the waste management chain. They are identified with their locally built two-wheeled traction vehicles used for the collection of MSW from homes with no access roads to MSW pick-up trucks. Fig. 4 shows the MSW management structure in Maputo.

6 CONCEPTUAL MODEL FOR EDUCATION AND OUTREACH ON MSW MANAGEMENT

The conceptual framework (Fig. 5) is to change the attitude and actions of the Mozambican people to achieve a sustainable waste management program. The proposed conceptual model on education and outreach on MSW management is a tool to enlighten the people of the society on waste disposal ethics, solid waste management benefits, and stakeholders' participation to attain sustainability in the waste management sector. This model is a composition of the four components, which include: principal actors, route, procedure, and people. The model is further expanded in Fig. 6 for proper understanding.

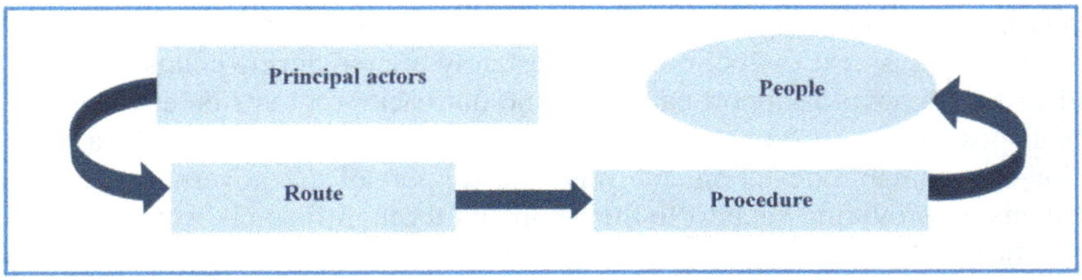

Figure 5: Conceptual framework.

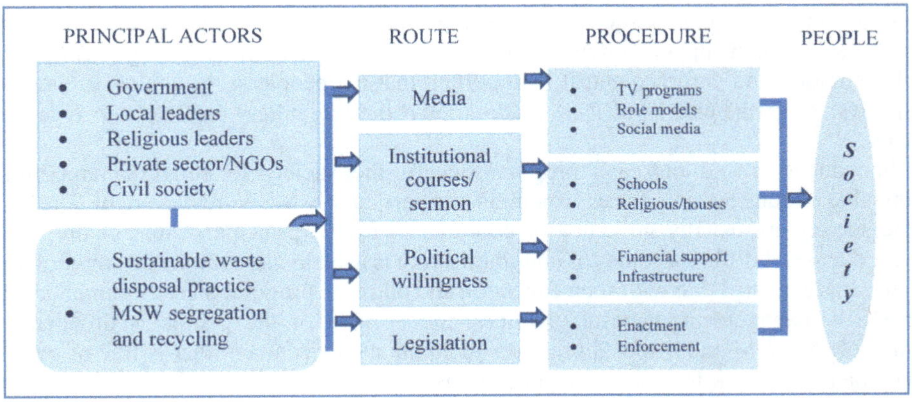

Figure 6: Education and outreach framework. *(Source: Adapted from [4].)*

Fig. 5 shows the main components of the education and outreach framework which include principal actors, route, procedure and people. The main objective of the principal actors in this study is to enlighten the people on environmental ethics and waste management practices and its benefits. The principal actors comprise of the government, local leaders, religious leaders, private sector, NGOs, as well as the civil society (see Fig. 6).

The principal actors convey their messages on and waste management practices with lots of emphasis on the benefits through some routes (see Fig. 6). This routes where the education and awareness are conveyed includes through the media, institutional courses/sermon, political willingness and legislation. The media remains one of the fastest ways of dissemination information to the public because people want to be well informed, learn, and perhaps satisfy their leisure time with pleasure on the current trends locally and globally. Also, information like is conveyed using different procedures with role models using their social media platforms to educate the people, demonstration or pronouncement in movies/ cartoons, as part of the lyrics in music, as radio adverts and broadcast and many more. Today, social networks play an important role in disseminating information to both urban and rural communities. Considering that most of the rural population is illiterate, music, radio broadcast and images on billboards is likely be the best way to disseminate information.

Furthermore, another way of educating the society on environmental ethics and waste management practices and its benefits is simply by introducing it as a subject/course are another in schools (elementary, high and tertiary). The recipient of such education in schools is likely to reflect in their perception and practice. This type of education can also be extended to orphanage homes, and religion houses (churches, mosques, temples) as sermon where the phrase "cleanliness is next to godliness" is constantly uttered during counselling.

Political willingness to support education and outreach programs on environmental ethic on waste disposal practices and waste management by the government will be a positivity for any nation's sustainable development. With the support of the government, there will be commitments by providing support to the campaign through financial support, employment of competent staff to educate the people, providing basic infrastructures and assured equality in services will be provided. For instance, MSW infrastructures and operation will not be focused only in the high-income districts, but it will rather be for all the inhabitants.

Again, the legislative arm of the government needs to see the importance to supporting environmental development concept, as weak policies and legislation have been detrimental to the environmental and the society. Hence, the existing legislation can leverage on or rather create new laws that supports environmental consciousness and waste management with full implementation and enforcement. It is important that the people in the society are aware of such laws. Nevertheless, it must be made known that defaulters are likely to face heavy penalty.

The education and outreach program benefit the society to enlighten and build a sustainable attitude towards the environment and particularly MSW management. The people of the society are usually governed or coordinated by the principal actors of the society. Hence, the use of different routes of communication is used to disseminate information to the society. It is expected that with such education and outreach program on environmental ethics and MSW management disposal practices, the horizon of the people is broadened on sustainable MSWM and disposal practices to attain an environment that is free of pollution and human health risk for the future generations.

7 CONCLUSION

In Mozambique, the weak policies and legislation of MSW management are mere pronouncement, non-implementation, and low enforcement, has contributed to the degradation of the urban environment. However, the creation of awareness and education on environmental ethics, and sustainability, enabling participatory management of MSW, to the people will take some time to produce the desired effect. A better environmentally educated society tends to understand that environmental awareness contributes to environmental sustainability and, consequently, it mitigates the human health risk and promote an aesthetic environment. This study has shown and discussed a conceptual model that emphasized that MSW management is not solely the responsibility of the government but a collective responsibility of all stakeholders to enhance the building of a sustainable environment. Hence, the outreach and education programs in MSW management should be a continuous and practicable exercise for sustainable development of the society. Also, it was deduced from the conceptual model that MSW management should begin with civil society and municipalities to establish an integrated solid waste management mechanism that are best suited to each of the capital cities and whole Mozambique. Again, synergy between principal actors and the society is likely to drive urban development of cities and towns.

In recommendation, the government should give priority to the environment by ensuring that solid waste management policies and legislation are revived, implemented and enforced in all arms of the government. The sensitization programs of solid waste management should not be resident in the cities alone but rather be extended to the rural communities continuously and not to be abolished by any change in government so that a good environmental consciousness is built and passed on to the future generation. The rate of urbanization in Mozambique requires adequate planning and efficient strategies in the management of the MSW generation, hence this study is likely to be the baseline line strength of such MSW management programs, especially with the channels of education, communication and participation practice.

ACKNOWLEDGEMENTS

Our appreciate goes to the Director of Natural Science and Mathematics, Universidade Pedagógica de Maputo, Maputo, Mozambique for his technical support and opportunities. Also, we truly appreciate the support of the Director of the Environmental Science and Technology Program, Faculty of Sciences, University of Porto, Portugal during this research.

REFERENCES

[1] Aderoju, O.M., Dias, G.A. & Gonçalves, A., A GIS-based analysis for sanitary landfill sites in Abuja, Nigeria. *Environment, Development and Sustainability*, 2018. https://doi.org/10.1007/s10668-018-0206-z.

[2] Nnaji, C. & Utsev, T., Climate change and solid waste management: A balanced assessment. *Journal of Sustainable Development in Africa*, **13**(7), pp. 17–34, 2011.

[3] Kadafa, A.A., Latifah, A.M., Abdullah, H.S. & Sulaiman, W.N.A., Current status of municipal solid waste practise in FCT, Abuja. *Research Journal of Environmental and Earth Sciences*, **5**(6), pp. 295–304, 2013.

[4] Aderoju, O.M., Dias, A.G. & Guimarães, R., Building an integrated perception and attitude towards municipal solid waste management in Nigeria. *Wastes 2015: Solutions, Treatments and Opportunities*, eds C. Vilarinho, F. Castelo & M. Russo, CRC Press, pp. 7–12, 2015.

[5] Jackson, T., *Motivating Sustainable Consumption. A Review of Evidence on Consumer Behaviour and Behavioural Change*, Sustainable Development Research Network (SDRN), 2005.

[6] Instituto Nacional de Estatística de Moçambique (INE), Apresentação dos Resutados do IV Recenseamento Geral da População, Censo de 2017. www.ine.gov.mz. Accessed on: 15 Dec. 2019.

[7] WHO, *Report on Environmental Sanitation in African Region; Trend and Analysis*, The 43rd Session of the World Health Organization Committee for Africa, Gaborone, Botswana, Sep. 1993, AFR/RC43/R2.

[8] IPCC, Solid waste disposal. *Guidelines for National Greenhouse Gas Inventories, Volume 5: Waste*, eds R. Pipatti & P. Svardal, P, Geneva, Switzerland, 2016.

[9] UNWSSD, *World Summit on Sustainable Development, Implementation Plan: Johannesburg Declaration Report*, United Nations Publications, Mar. 2002. www.un.org.esa/sustdev/documents/johannesburg%20Declaration. Accessed on: 14 Dec. 2019.

[10] Schübeler, P., *Conceptual Framework for Municipal Solid Waste Management in Low-Income Countries*, eds K. Wehrle & J. Christen, UNDP/UNCHS/World Bank/SDC Collaborative Programme on Municipal Solid Waste Management in Low-Income Countries: St. Gallen, Switzerland, 1996.

[11] Rousta, K. & Dahlén, L., Source separation of household waste, technology and social aspects. Resource Recovery to Approach Zero Municipal Waste, eds M.J. Taherzadeh & T. Richards, Boca Raton: CRC Press, pp. 61–77, 2015.

[12] Sallwey, J., Hettiarachchi, H. & Hülsmann, S., Challenges and opportunities in municipal solid waste management in Mozambique: A review in the light of nexus thinking. *AIMS Environmental Science*, **4**(5), pp. 621–639, 2017. DOI: 10.3934/environsci.2017.5.621.

[13] Ferrari, K., Gamberini, R. & Rimini, B., The waste hierarchy: A strategic, tactical and operational approach for developing countries – The case study of Mozambique. *Int. J. Sustainable Dev. Plan*, **11**, pp. 759–770, 2016.

[14] Muchangos, A., Moçambique, Paisagens e Regiões Naturais, Tipografia Globo, 1999. 01048/FBM/93, República de Moçambique, Maputo.

[15] Karani, P. & Da Costa, J., The role of knowledge in improving waste management for sustainable development: The case of Maputo Mozambique, 2007. www.environmentalexpert.com/Files/0/articles/16550/MozambiqueWasteManagement--PatrickKarani--article2.doc. Accessed on: 18 Nov. 2019.

[16] Tas, A. & Belon, A., A comprehensive review of the municipal solid waste sector in Mozambique, 2014. www.associacao-mocambicana-reciclagem.org. Accessed on: 18 Nov. 2019.

[17] MITADER (Ministério da Terra, Ambiente e Desenvolvimento Rural), Statistical data of municipal waste management in major cities of Mozambique, 2019. www.ruralmoc.gov.mz/. Accessed on: 12 Dec. 2019.

[18] AMOR, Project Catalogue, 2013. www.associacao-mocambicana-reciclagem.org. Accessed on: 21 Nov. 2019.

[19] Buque, L.I.B. & Ribeiro, H., Overview of the selective waste collection with pickers in Maputo Municipality, Mozambique: Challenges and Perspectives. *Saúde Soc.* **24**, pp. 298–307, 2015.

[20] Decree No. 94/2014, Regulation on Municipal Waste Management, Maputo, 31 Dec. 2014.

[21] Resolution No. 86/AM/2008, Posture of Cleaning of Solid Municipal Waste in the Municipality of Maputo, Maputo, 22 May 2006.

[22] Stretz, J., Economic instruments in solid waste management. Eschborn, Germany: Deutsche Gesellschaft für Internationale Zusammenarbeit (GIZ) GmbH, 2012. www.giz.de/de/downloads/giz2012-en-economic-instruments-mozambique.pdf.

[23] Federal Ministry for Economic Cooperation and Development, Solid Waste Management Master Plan (SWMMP). Maputo, 2008. www.giz.de/en/downloads/giz2012-en-economic-instruments-mozambique.pdf.

PROPOSAL FOR SOCIAL INDICATORS TO IMPROVE MUNICIPAL SOLID WASTE MANAGEMENT: A PERUVIAN CASE STUDY

LUIS IZQUIERDO-HORNA, MIKER DAMAZO & DEYVIS YANAYACO
Centro de Investigación de Riesgo, Resiliencia y Cambio Climático, Universidad Tecnológica del Perú, Peru

ABSTRACT

The accumulation of waste in Peru has become a problem that affects the entire population and has generated the need to understand how and why this situation has developed. This situation can be described and understood through an examination of the most relevant social indicators of a very specific community. Consequently, it is necessary to understand the link between the situational reality of solid waste accumulation and the social indicators that are relevant to this process. Therefore, the main objective of this research is to contribute to the improvement of municipal solid waste management plans, specifically solid waste collection tasks. To this end, a methodological tool that focuses on the study of social indicators (e.g., age, education level, and financial income, among others) is proposed to identify the sectors most prone to solid waste accumulation. This study proposes a set of indicators that describe the economic, social, and environmental conditions of the location inhabitants and answers the questions initially posed. The district chosen for this case study is Puente Piedra. This research resulted in a set of social indicators divided into sociodemographic, socioeconomic, and sociocultural categories that can be used to forecast future tasks in sectors prone to solid waste accumulation.

Keywords: solid waste, social indicators, waste management, Peru.

1 INTRODUCTION

Solid waste generation has become a global problem because the planet has experienced alarming levels of production [1]. According to Kaza et al. [2], the world generated 2.01 billion tons of municipal solid waste in 2016 and this number is expected to increase to 3.40 billion tons by 2050. Consequently, it is important to manage it efficiently, as well as raise awareness among populations to control the problem that is becoming more acute daily. The elimination of solid waste has undoubtedly become a problem that must be treated with the utmost urgency [3].

On the other hand, factors such as continuous economic growth, urbanization, and industrialization, hinder the correct management of solid waste and create a problem for the authorities with responsibility for executing it efficiently [4]. One of the major sources of environmental pollution is domestic solid waste, especially in developing countries, where only 4% of the waste generated is recycled [2]. Therefore, efficient and modern solutions are being sought in different regions of the world to minimize the production of solid waste; however, the efficacy and success achieved have been uneven and limited [2].

Against this background, this research focuses on studying the process of solid waste generation/accumulation, considering solid waste as any substance or product in a solid or semi-solid state that when it has fulfilled its function or been used and/or served an activity, then is subsequently discarded. The mismanagement of solid waste can create a risk to health and the environment [5], [6]. On the other hand, the management process is divided into hazardous and non-hazardous waste and classified as municipal and non-municipal waste accordingly [7]. The main actors in solid waste policy are the competent authorities, companies, and the population, the latter being the most decisive [8].

www.witpress.com, ISSN 1743-3541 (on-line)
doi:10.2495/WM200061

For this reason, the National Environmental Action Program and the National Environmental Action Plan in Peru have agreed on a priority objective: the comprehensive management of solid waste at the national level. Established goals must be met by 2021, including environmental education about responsible consumption [9]. From this perspective, the Ministry of Environment (MINAM) has identified 28 critical points in 43 districts of Lima. In addition, the Environmental Assessment and Inspection Agency (OEFA) has determined 148 critical points of waste generation (also in Lima), which add up to a total of 8,670 tons of waste per day [10]. However, of the total national waste generated, only 43.5% is deposited in sanitary landfills and only 55.4% is deposited correctly or safely [11]. According to Alata [12], the Metropolitan Municipality of Lima registered the generation of more than three million tons of garbage in 2017, which is equivalent to more than 8,970 tons per day. This is an uncomfortable situation for citizens as evidenced by a survey conducted in Lima. The survey results reflected citizens' dissatisfaction with the waste collection service (32.6%) and the lack of an effective and efficient recycling system (18.9%) [13].

At an international level, the situation is not very different from that in Peru. A series of projects are being planned and implemented to improve solid waste management. For example, in Colombia, a methodological proposal was developed to determine how much waste of residential origin is generated in developing countries. The results obtained were acceptable for medium-sized cities. The research team conducted prior visits and home surveys which led to a fluid and organized collaboration with the community itself [14]. In Iran on the other hand, the same strategy was applied to 1,782 households to determine the factors associated with domestic solid waste generation, and information was collected through a questionnaire that evaluated sociodemographic factors (e.g., age, marital status, gender, etc.) [15]. Likewise, in Canada, Kannangara et al. [16] conducted research to develop a predictive model based on demographic and socioeconomic indicators. This model was created through a residential mapping exercise in 220 municipalities. They analyzed the data collected detailing the generation of annual residential waste and obtained successful results with a good predictive percentage (72%).

In the same way, Torres et al. [17] developed a comprehensive management model in Mexico that addressed the problem of solid waste accumulation in public places. The study highlighted the importance of people as fundamental change agents. Similarly, in the Netherlands, an estimate of the amount of solid food waste was made through a survey of 763 citizens using a cell phone application and classification techniques of household solid waste in 130 households [18]. A similar project was conducted in Dehradun, India, where the quantity and quality of household waste were studied in terms of socioeconomic groups and family size [19]. In addition, a study was conducted in Cape Haitian, Republic of Haiti, through a survey of socioeconomic groups and waste generation [20].

An important feature of the studies shown is that, despite their variety, the methods applied in different countries have in many cases involved creative or traditional strategies, such as surveys. In this sense, this research hopes to establish a tool to develop knowledge about the accumulation of solid waste and to elucidate the social, economic, and cultural characteristics that influence this process to improve citizens' living conditions.

This research developed a set of social indicators that identify key spots of solid waste generation in urban areas. Studies conducted from this perspective can be used to geo-reference the relationship between selected social indicators and solid waste generation. This article is divided into five sections. The first section focuses on describing a global vision of the subject as an introduction. The section that follows describes the methodology and important characteristics of the case study (Puente Piedra district). Section 3 describes the main results obtained. Section 4 discusses the most relevant results and finally, the Section 5 presents a conclusion.

2 METHODOLOGY

In this section, a description of the methodology, the case study, and the information sources used for this research are presented. The study focuses on an urban area of Lima, in Puente Piedra district. The methodological framework is presented in Fig. 1 and explained in the following paragraphs. The first part details the information collection process and the selection of the case study, using information from work conducted by the Ministry of Environment of Peru (MINAM) and the National Institute of Statistics and Informatics of Peru (INEI). The second part identifies the social indicators used based on the characteristics of the study area and relevant prior information; and finally, an analysis was conducted and selected indicators proposed.

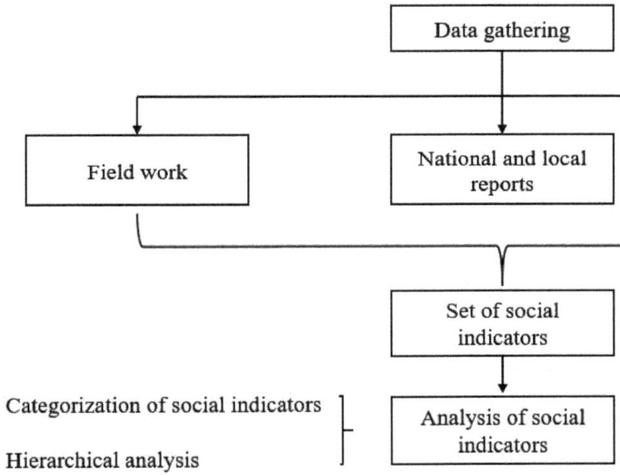

Figure 1: Methodological framework.

2.1 Description of the case study

Puente Piedra is located north of Lima Province and is a district with a total area of 72.81 km^2 and a population of 329,675 inhabitants [21]. It was chosen not only for its accessibility but also because it is one of the most contaminated districts in Peru [22]. Given the district's land area, we decided to focus on two sectors: "Santa Juana de Copacabana" and "Los Frutales del Norte I" Associations. The location and spatial features of the chosen sectors are presented in Fig. 2.

2.2 Identification of social indicators

According to Acuña and Valera [23], one community's social indicators do not necessarily explain the behavior of wider society; therefore, to have greater control and minimize errors, we aimed to identify social indicators specific to the case study location to identify the

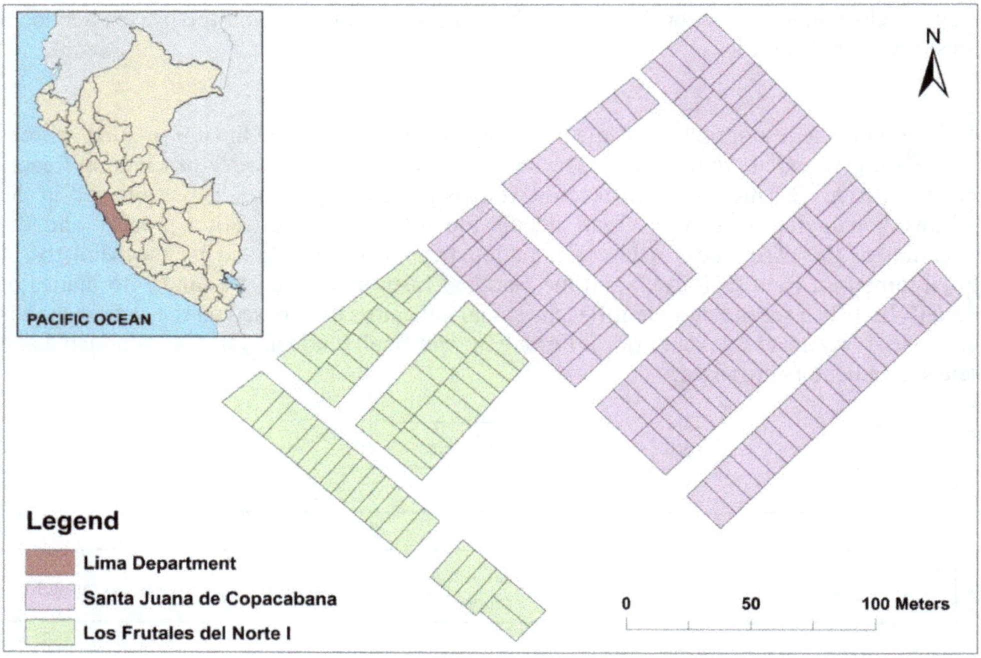

Figure 2: Case Study: Santa Juana de Copacabana and Frutales del Norte I.

average behavior of residents of the study area. For this reason, it was necessary to define what to measure and how to measure it. This project analyzed the most relevant social characteristics influencing the solid waste accumulation process. Therefore, we validated the features of each social indicator to determine its relevance and how representative it was of the topic being studied [24], [25].

We also needed to analyze demographic data and collect information samples to design a survey that covered all the study indicators [19]. For this project, the residents of the study area were interviewed and this information was used to identify the relevant social indicators. This selection process involved establishing communication strategies with the residents to obtain their greatest possible collaboration including awareness workshops, prior visits, and interviews. Another important thing to consider was that each indicator needed linking to both the locality and the dwellings in an environmental context, and the intrinsic characteristics of the physical environment [26].

Social indicators were selected using an existing bibliography and expert judgment. The selection process involved interviews with the residents of both locations, to identify correlations between existing living conditions and the generation of infectious foci. A total of 176 residents were interviewed. Therefore, the assessment process and results obtained in this study are only applicable to a certain time and space; however, the methodology is reproducible at all levels of scale. Finally, the selected social indicators were grouped into three categories for clarity: sociodemographic, socioeconomic, and sociocultural factors. Table 1 presents the set of social indicators proposed for subsequent application in the study area, to make it possible to assess the correlation between the social profile obtained and the locations of solid waste accumulation.

Table 1: Classification of indicators according to validity and representation.

Categories	Social indicators
Sociodemographic conditions	Age
	Education level
	Population density
	Land use
Socioeconomic conditions	Consumption pattern
	Existence of ambulatory trading
	Average monthly income
Sociocultural conditions	Local customs
	Waste management awareness
	Participation in training

2.3 Analysis of social indicators

This section explains the procedure involved in applying the social indicators previously selected for identifying possible sectors of solid waste accumulation. For this, the study carried out by Thomas Bohorquez [27] was taken as a reference, since it used a mixture of analysis and data processing tools to improve city planning. The result of this adaptation is shown in Table 2. The sociodemographic, socioeconomic, and sociocultural profile of the study area was expected to be the final product. Multivariate analysis can be conducted to obtain subsequent geo-referencing and prediction.

3 RESULTS

The main contribution of this methodology was a set of ten social indicators characterized in three dimensions: sociodemographic, socioeconomic, and sociocultural.

The sociodemographic dimension comprised the following descriptors: age, education level, population density, and land use. This category will permit a better understanding of whether there is any relationship between its descriptors and solid waste accumulation such as how relevant education level or age is to these processes. On the other hand, the socioeconomic dimension comprised the following descriptors: consumption patterns, the existence of ambulatory trading, and average monthly income. This layer of information provides an understanding of the economic profile of the study sector and its relationship with solid waste accumulation. For example, if the consumption pattern occurs daily rather than weekly, it could be that the generation of solid waste is greater. Finally, the sociocultural dimension comprised the following descriptors: local custom, awareness of solid waste management policies, and participation in training. This last level of information permits a better understanding of the social constructs in the sector analyzed. For example, it will describe residents' behavior patterns and incentives to take the garbage out (waiting for the collector or leaving it in a place that serves as an accumulation spot).

The ten descriptors mentioned above provide an understanding of the population in action, thereby helping to protect their environment by identifying whether there is a lack of opportunity for development to improve residents' living conditions. To understand the importance of the proposed indicators, official information provided by the INEI during the last 2017 census was used because it collected multiple social indicators to assess inhabitants living conditions. Statistics for the study area indicated that 82% of dwellings were inhabited. Another important fact is that 71.3% of dwellings have both basic water and electricity

Table 2: Definition and categorization of indicators.

Indicator	Definition	Descriptors	Contribution to the description of critical spots
Age	Age range of the person	(4) 15 to 25 years (3) 26 to 35 years (2) 36 to 45 years (1) 46 years old and over	Determines the predominant age range of the study population
Education	Acquired level of education	(4) None (3) Primary school (2) Secondary school (1) University/technical school	Has inference with the level of educational attainment
Land use	Anthropic activity at the property	(3) Business (2) Multifamily housing (1) Single-family housing	Associates waste generated by the family with waste disaggregated in open spots
Population density	Number of persons per household	(4) 10 or more (3) 7 to 9 (2) 5 to 6 (1) 1 to 4	Associates the amount of waste generated with the number of persons per household
Average monthly income	Household monthly income per person (in soles)	(4) S/3001 or more (3) S/2001–S/3000 (2) S/1001–S/2000 (1) S/0–S/1000	Associates the purchasing power (consumption) of each household with the generation of waste
Consumption patterns	Household grocery shopping frequency	(4) Daily (3) Weekly (2) Biweekly (1) Monthly	Refers to consumption frequency and accumulated waste

Table 2: Continued.

Indicator	Definition	Descriptors	Contribution to the description of critical spots
Existence of ambulatory trading	Existence or non-existence of informal activities near the household	(2) No (1) Yes	Generation of inappropriate conditions leading to waste accumulation
Local customs	Routine activities for solid waste disposal	(2) I leave the garbage at the door of my house (1) I wait for the garbage collector to arrive	Affects the frequency of waste disposal in open spots
Awareness of solid waste management policies	Awareness of the existence of raising awareness campaigns	(2) No (1) Yes	Knowledge to manage waste contamination problems
Participating in training workshops	Training and degree of awareness of environmental care	(2) No (1) Yes	Identifies the level of participation and knowledge of solid waste management.

services. In terms of educational level, 44.4% of the population had completed secondary school and 4% never had an education. Age stratification measures indicated that 46.6% of the residents were between 18–44 years old. Economic profiling indicated that 34.35% were independent workers, 32.07% were employees and 23.84% were engaged in mixed activities. Finally, the family composition tended in most cases to consist of 4 people (70.02%) [28].

4 DISCUSSION

The proposed set of indicators provides a robust and replicable tool to identify places prone to solid waste accumulation. In addition, it could contribute to the optimization of municipal management, providing methods that make it possible to understand residents better and implement a management plan according to their needs. It is important to understand the attitudes and beliefs of a population to understand how individual behavior impacts on environmental contamination [29]. Therefore, social indicators help identify the behavior that generates solid waste accumulation spots, to better understand this phenomenon and implement corrective measures to improve waste treatment leading to an overall reduction in waste dumping.

Solid waste accumulation areas are difficult to predict using a single social descriptor; therefore, it was necessary to combine different descriptors grouped into the categories that we have addressed: sociodemographic, socioeconomic, and sociocultural factors. Given that it was important for the authors to prioritize indicators in order of the relevance to their respective categories, we produced a hierarchical weighting matrix based on Saaty's studies [30]. Table 3 presents the results of the hierarchical analysis.

Table 3: Hierarchical analysis of social indicators.

Categories	Social indicators
Sociodemographic conditions (48%)	Age (6.6%)
	Education level (42.2%)
	Population density (38.1%)
	Land use (13.1%)
Socioeconomic conditions (11.5%)	Consumption pattern (40.5%)
	Existence of ambulatory trading (11.5%)
	Average monthly income (48%)
Sociocultural conditions (40.5%)	Local customs (55.7%)
	Waste management awareness (32%)
	Participation in training (12.3%)

When evaluating the relevance of each category, we observed that sociodemographic and sociocultural indicators were the most relevant with 88.5% since local residents' customary disposal of solid waste depends on these. However, sociodemographic indicators including education level, closely followed by population density, and the number of household occupants directly affects the generation of waste. In the area chosen, age was not relevant, since the vast majority of the population were young [28]. In the same way, at the sociocultural level, local customs were a more relevant indicator than awareness and participation in issues related to solid waste management. This organizational hierarchy indicates that social constructs are inherent in the formation of individual behavior. Finally, socioeconomic indicators revealed that the average monthly income and consumption patterns go hand in hand.

5 CONCLUSIONS

This research identified a set of social indicators that influence solid waste accumulation in urban areas. Three categories of socioeconomic, sociocultural, and sociodemographic factors were used to bundle the proposed set of indicators. Sociodemographic indicators describe the resident's social profile: land use, age, population density, and education. Socioeconomic indicators describe the relevant economic parameters influencing the accumulation process: average monthly income, consumption patterns, and the presence of ambulatory trading. Meanwhile, sociocultural indicators describe residents' awareness and subsequent behavior. Finally, this research is presented as a starting point for future research papers which will implement and validate the descriptors analyzed in this proposal.

ACKNOWLEDGEMENTS

This project was funded by Universidad Tecnológica del Perú (UTP), within the framework of the "Research Projects I+D+i 2019" agreement. The authors would like to thank Yuriko Saito and Ramzy Kahhat for their valuable comments on the previous version of this manuscript.

REFERENCES

[1] Hoornweg, D. & Bhada-Tata, P., What a waste: A global review of solid waste management. Urban Development Series Knowledge Papers, Washington, 2012. https://openknowledge.worldbank.org/handle/10986/17388.

[2] Kaza, S., Yao, L., Bhada-Tata, P. & Van Woerden, F., What a waste 2.0: A global snapshot of solid waste management to 2050. Urban Development Series, Washington, 2018. http://hdl.handle.net/10986/30317.

[3] Wang, L., Yan, D., Xiong, Y. & Zhou, L., A review of the challenges and application of public–private partnership model in Chinese garbage disposal industry. *J. Clean. Prod.*, **230**, pp. 219–229, 2019. www.sciencedirect.com/science/article/pii/S0959652619315446.

[4] PNUMA, Developing an integrated solid waste management plan: training manual. Kusatsu: UNEP, 25 pp., 2009. www.pseau.org/outils/ouvrages/ietc_developing_integrated_solid_waste_management_plan_training_manual_vol_2_assessment_of_current_waste_management_system_and_gaps_therein_2009.pdf.

[5] Espinoza, C., Management of domestic and non-domestic solid waste in the municipal management of the city of Huancavelica, period 2016. Repositorio Institucional (UNH), Universidad Nacional de Huancavelica, 2018. http://repositorio.unh.edu.pe/handle/UNH/1830.

[6] Salazar de la Rosa, N., Solid waste management in food companies. Universidad Nacional Agraria La Molina, 2018. http://repositorio.lamolina.edu.pe/handle/UNALM/3586.

[7] MINAM, Legislative Decree No. 1278. R.M.N° 024-2017-VIVIENDA Perú, 35 pp., 2017.

[8] INEI, *Peru: Environmental Statistics Yearbook 2014*, Lima, 574 pp., 2015. www.inei.gob.pe/media/MenuRecursivo/publicaciones_digitales/Est/Lib1197/libro.pdf.

[9] MINAM, National plan for the comprehensive management of solid waste 2016–2024. SINIA: Perú, p. 85, 2016. https://sinia.minam.gob.pe/documentos/plan-nacional-gestion-integral-residuos-solidos-2016-2024.

[10] OEFA, Environmental monitoring of solid waste from provincial municipal management. SINIA: Lima, 2014. https://sinia.minam.gob.pe/documentos/fiscalizacion-ambiental-residuos-solidos-gestion-municipal-provincial.
[11] Cárdenas, G., The city and the garbage. RPP, 2018. https://rpp.pe/lima/actualidad/la-ciudad-y-la-basura-lima-retrocede-en-la-lucha-por-controlar-sus-residuos-solidos-noticia-1114285.
[12] Alata, P., Sustainable country, GRUPO RPP: Lima, 2019.
[13] Lima Cómo Vamos, IX Perception report on quality of life in Lima and Callao. Encuesta Lima como vamos, Lima, 2018. www.limacomovamos.org/cm/wp-content/uploads/2018/12/EncuestaLimaComoVamos2018.pdf.
[14] Klinger, R., Olaya, J., Marmolejo, L. & Madera, C., Sampling plan for the quantification of residential solid waste generated in urban areas of intermediate size cities. *Rev. Fac. Ing. Univ. Antioquia.*, **48**, pp. 76–86, 2009. www.scielo.org.co/scielo.php?script=sci_abstract&pid=S0120-62302009000200008&lng=e&nrm=iso&tlng=e.
[15] Pakpour, A.H., Zeidi, I.M., Emamjomeh, M.M., Asefzadeh, S. & Pearson, H., Household waste behaviors among a community sample in Iran: An application of the theory of planned behavior. *Waste Manag.*, **34**(6), pp. 986–986, 2013. https://linkinghub.elsevier.com/retrieve/pii/S0956053X1300514X.
[16] Kannangara, M., Dua, R., Ahmadi, L. & Bensebaa, F., Modeling and prediction of regional municipal solid waste generation and diversion in Canada using machine learning approaches. *Waste Manag.*, **74**, pp. 3–15, 2018.
[17] Torres, A., Garnica, J. & Niccolas, H., Methodology to generate a comprehensive solid waste management model. Pachuca case study. *XV International Congress of Research in Administrative Sciences*, 2011.
[18] Van Dooren, C., Janmaat, O., Snoek, J. & Schrijnen, M., Measuring food waste in Dutch households: A synthesis of three studies. *Waste Manag.*, **94**, pp. 153–164, 2019.
[19] Suthar, S. & Singh, P., Household solid waste generation and composition in different family size and socioeconomic groups: A case study. *Sustain. Cities Soc.*, **14**(1), pp. 56–63, 2015. www.sciencedirect.com/science/article/abs/pii/S2210670714000821.
[20] Philippe, F. & Culot, M., Household solid waste generation and characteristics in Cape Haitian city, Republic of Haiti. *Resour. Conserv. Recycl.*, **54**(2), pp. 73–78, 2009. www.sciencedirect.com/science/article/abs/pii/S0921344909001359.
[21] Torres, A., Del Castillo, A., Mezones, J., Yong, J. & Revichagua, C., Statistical Compendium Lima Province 2018. www.inei.gob.pe/media/MenuRecursivo/publicaciones_digitales/Est/Lib1655/libro.pdf.
[22] MINAM, Ministry of Environment identifies 92 districts that need to take action to improve solid waste management, 2018. www.minam.gob.pe/notas-de-prensa/ministerio-del-ambiente-identifica-92-distritos-en-situacion-de-riesgo-por-manejo-de-residuos-solidos/.
[23] Acuña, S. & Valera, V., Indicators of waste generation and solid waste in Venezuela. *Redisa*, **1**, pp. 1–11, 2008.
[24] Izquierdo-Horna, L. & Kahhat, R., Methodological framework to integrate social and physical vulnerability in the prevention of seismic risk. *WIT Transactions on Engineering Sciences*, vol. 121, WIT Press: Southampton and Boston, pp. 69–79, 2018.

[25] Ibañez Forés, V., Bovea, M.D. & Coutinho-Nóbrega, C., Proposal of social indicators to assess the social performance of waste management systems in developing countries: A brazilian case study. *Perspectives on Social LCA: Contributions from the 6th International Conference*, Paraíba, pp. 95–103, 2020. www.researchgate.net/publication/335612756_Proposal_of_Social_Indicators_to_Assess_the_Social_Performance_of_Waste_Management_Systems_in_Developing_Countries_A_Brazilian_Case_Study.
[26] Durán, C., Spatial analysis of social, economic, physical and environmental vulnerability conditions in the Colombian territory. *Perspect. Geográfica*, **22**(1), pp. 11–32, 2017. www.scielo.org.co/pdf/pgeo/v22n1/0123-3769-pgeo-22-01-00011.pdf.
[27] Thomas Bohorquez, J., Social vulnerability assessment of natural hazards in Manzanillo (Colima). A methodological contribution. *Investig. Geogr.*, 81(100), pp. 79–93, 2012. www.scielo.org.mx/scielo.php?script=sci_arttext&pid=S0188-46112013000200007.
[28] INEI, National Census 2017: XII of population,VII of housing and III of indigenous communities, 2017. https://censos2017.inei.gob.pe/redatam/.
[29] Brown, D.P., Garbage: How population, landmass, and development interact with culture in the production of waste. *Resour. Conserv. Recycl.*, 98, pp. 41–54, 2015. www.sciencedirect.com/science/article/pii/S0921344915000440.
[30] Saaty, T., *The Analytic Hierarchy Process*, New York: McGraw-Hill, 1980.

COMPOSTING STRATEGY FOR DEVELOPING CITIES: A CASE STUDY OF BEIRA, MOZAMBIQUE

DARIO GUIRRERI[1], SILVIA SILVESTRI[2], ANDREA CRISTOFORETTI[2], ISACCO RAMA[3], RAMONA GIUREA[1], ELENA MAGARIL[4] & ELENA CRISTINA RADA[5]
[1]Environmental and Mechanical Engineering, University of Trento, Department of Civil, Italy
[2]Foundation Edmund Mach, Italy
[3]Foundation CAM – Consortium of Associations with Mozambique, Italy
[4]Department of Environmental Economics, Ural Federal University, Russian Federation
[5]Department of Theoretical and Applied Science, University of Insubria, Italy

ABSTRACT
This article aims to provide a contribution to the value and possibility of using composting as a tool for the treatment and management of the organic fraction of urban solid waste in developing contexts – specifically, the city of Beira, Mozambique. The aforementioned process should be intended not as an exhaustive tool but rather as a useful form of treatment to be employed within an Integrated Waste Management Plan. The high and diversified presence of materials suitable for the specific treatment in question is ascertained while also highlighting the diversification of the users that can be involved. Mixing strategies are drawn up in order to provide indications for a correct composition of the matrix that is intended to be started as a process and, according to the quantities considered (according to a modular approach), the production that cán derive from it. A technical proposal is then drawn up on the functional areas making up the plant in order to develop the process in question. In addition to promoting the composting process itself, the article aims to valorize its final product, the compost, which in certain specific contexts, such as the one considered, can be useful or even necessary.
Keywords: composting, developing countries, Mozambique, MSW management, organic waste.

1 INTRODUCTION

Waste Management is the second most important problem after water quality in developing countries and worldwide [1]–[8]. According to the United Nations Environment Programme (UNEP) [9] waste disposal is a relevant problem in the less developed countries due to the demographic increase, poverty and high rate of urbanization, all that combined with inefficiency and under-financing by Governments in providing an adequate strategy for the management of municipal solid waste (MSW) [10]–[14].

Factors such as waste composition, technologies and/or lack of infrastructures, have been recognized as basic elements to set up a good and sustainable MSW management in developed and developing countries [15]–[18]. Generally, in the developing countries the only form of final disposal is generally open landfills, which periodically develops uncontrolled combustion with the inevitable production of atmospheric pollution and leachate [19]–[25].

MSW generated in developing countries is characterized by a high density and high content of organic matter, few metals and plastics content [26]–[30]. Big problems refer to the inefficient facilities for MSW collection, low treatment technologies and presence of slums built in a random and unplanned manner with unavoidable problems of accessibility [31]–[35].

At national level, in Mozambique, the Decree No. 13/2006 introduces Waste Management Plans (WMP) and includes relevant rules, provides principles, objectives, tools and guidelines for an Integrated Solid Waste Management (ISWM) [36].

The city of Beira (Conselho Municipal da Beira – CMB), is making concrete efforts in improving provided services to citizens promoting an international partnership coordinated

WIT Transactions on Ecology and the Environment, Vol 247, © 2020 WIT Press
www.witpress.com, ISSN 1743-3541 (on-line)
doi:10.2495/WM200071

by Consortium of Associations with Mozambique (CAM). CMB has drafted the "*Postura Municipal para a Gestão de Resíduos Sólidos*", which aims not only to find management tools, but also to improve life conditions of population, in order to establish the sufficient conditions for a sustainable development.

The use of *compost* as a strategy for the management of the organic fraction of municipal solid waste (OFMSW), in Mozambique, it was already used in Maxixe [37]. Also, in Beira, composting has been identified as a possible method of OFMSW treatment, with a view to valorisation and reuse [38]. This article presents the results and proposals for the Beira case.

2 MATERIALS AND METHODS

Beira is the capital of the Sofala Province and the second city of Mozambique with a population of about half million of inhabitants. It represents the second industrial park and the third biggest seaport in Mozambique for the import-export flows. The city's economy is mainly based on trade and port activity and the commercial and transport sectors appear to be the most developed; a sector of equal importance is certainly represented by agriculture.

The MSW management is on the responsibility of the VGUE (*Vereação de Gestão Urbana e Equipamentos*). The current state presents several problems both with regard to the equipment available, their excessive use as well as an incomplete coverage of the service for the areas of the city. A weak link in the chain is precisely the collection service. There are three different systems forms of waste collection: *contendores* (6 m^3), *tambores* (200 L) and direct arrangement on the ground. Container trucks, auto-compacting machines and tractors are the means used for the MSW collection, whose number of units is reduced, and the availability often not guaranteed. The final point of waste delivery, the municipal landfill, is not delimited and presents significant problems with regard to the access especially during the rainy period. The waste disposal is completely uncontrolled, so there are problems of both volumes and run-off; the landfill, as common in other developing contexts, presents the informal collection phenomena.

For the development of the present research, six sectors of different types of waste were considered (domestic, market waste, industrial, commercial, green street sweeping waste). The total daily production being about 523 t/d (190,822 t/y). The domestic sector represents the major contribution, 65%, followed by the market waste sector, 20% [38], [39].

In the context of Beira, different directions were taken that include a direct approach with local institutions, with managers and activities aimed at the quality-quantitative characterization of materials and/or the identification of spaces in order to combine all useful and necessary information to draft the project proposal for MSW management. The field work was divided into: identification of the matrices to be treated and their production points, transfer modes of the selected matrices, treatment plant, composting process and compost, intended destination.

Regarding the market and food wastes, data were collected together with VGUE. The percentage of organic matter (100 m^3/month) produced in these markets was identified through field analysis, direct comparisons with local responsibles and measurements of the apparent density of the waste were carried out in order to extrapolate the weight value of the effective quantity of organic matter produced. The characterization of waste from restaurants activities was conducted through questionnaires. The georeference of the restaurants was made in order to estimate the produced quantities of waste and its percentage of organic fraction (OF).

Thanks to the collaboration with VAPPMA (*Vereação de Agro-Pecuária, Pescas e Meio Ambiente*), and local responsibles, it was possible to retrieve data regarding the maintenance of the urban green waste (number of cropper trees), the pruned roads and the specifications relating to the operating conditions of the activities themselves (number of means used and their capacity). The average timber productions associated with a tree has been identified and has allowed the extrapolation of the average monthly production and maps have been drawn up to identify the urban areas undergoing maintenance.

Regarding seeping and sludge waste, data were obtained thanks to the collaboration with SASB (*Serviços Autónomos de Saneamento da Beira*) and with ETAR (*Estação de Tratamento de Águas Residuais*). The investigative activity concerning this type of material was possible thanks to a comparison with the treatment plant manager, the direct vision of operations conducted inside the wastewater treatment station and the estimation of the apparent density of the sludge in order to extrapolate quantitative production at monthly scale. Furthermore, chemical analyses have been carried out on the sludge, which has ascertained the suitability for use in the composting process.

The proposed plant, for the above considered materials, is shown in Fig. 1. Compost market opportunities represent a crucial point in order to understand the potential use of the final product. The problem does not consist in identifying the intended use, but in the cost-benefit aspect, in order to make the entire supply chain economically sustainable.

Chemical analyses have been taken on soil samples of some countryside in the outlying areas of the city and investigation in theoretical form have been carried out additionally. As regard the cost-benefit analysis, realities possibly interested in the purchase of compost have been investigated and are worthy of further investigation. The first stage was the identification and selection of the available matrices that it was decided to use in the composting treatment (their quality-quantitative characterization being a nodal point).

In this paper the OFMSW considered to be sent in the composting plant was divided into two components: market and catering waste. Regarding the first one, the production of the two main urban markets of the city, *Maquinino* and *Goto*, have been evaluated.

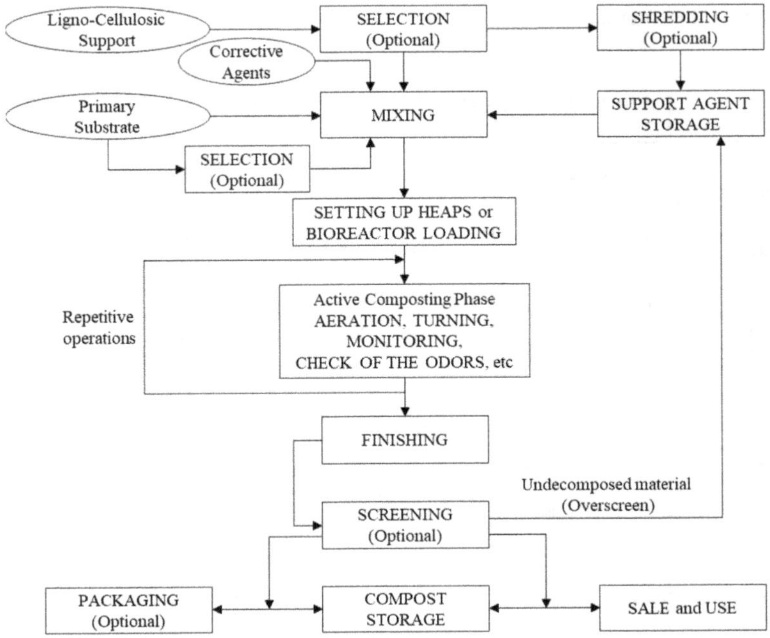

Figure 1: Flow chart of a composting plant.

3 RESULTS

Thanks to the collaboration with local bodies and VGUE, VAPPMA, SASB, ETAR, CAM and CMB, it was possible to obtained the determination of mass quantities, reported in Table 1, useful and necessary to draft the mixing plans.

The values reported have been time scaled, starting from the determination of the OFMSW production on daily basis (day of market) and considering the number of days of market along the year. Half of the containers considered in each market are also improperly used by citizens, so only those for use almost exclusively in the markets were taken into consideration: almost exclusively OFMSW was found with a low percentage of other materials (15%).

Regarding the restaurant activities, a map was created in order to have a graphical representation of their spatial arrangement within the urban fabric. Among those surveyed, 20, only half were taken in consideration. These are the major ones which presented at the same time a greater interest and willingness to collaborate. In Table 2, data regarding the OFMSW for the considered matrices are reported. The OFMSW quantities from these have a contribution of an order of magnitude lower than that of the markets but not less important for this reason. The reported data demonstrate also that it is essential to integrate and involve the users.

Regarding the lignocellulosic material to be used in the composition of the initial matrix, the green waste production was estimated considering the units undergoing maintenance, and the potential of the management service for urban green areas. A production range of 0.4–1.4 t/tree has been assumed, considering the size and variability of the tree type. The average production value of each tree was considered 0,9 t. In Table 3 the production fluctuation during a year is reported.

In order to verify the values, the calculations were performed according to the number of working days (visible in Table 4), the number of means used daily (during maintenance operations) and their capacity.

Table 1: Production of potential organic matter (OM).

Marketplace	Total annual production OM		Total monthly production OM	
	(m^3/y)	(t/y)	(m^3/month)	(t/month)
Maquinino	5,875.2	2,908.2	489.6	242.4
Goto	2,937.6	1,454.1	244.8	121.2
Total	8,812.8	4,362.3	734.4	363.5

Table 2: Quantitative production.

Source	Total monthly production O.M.
	(t/month)
Market waste	364
Restaurants	11

www.witpress.com, ISSN 1743-3541 (on-line)

Table 3: Monthly lignocellulosic material production.

Months	Pruned trees	Average production (t/month)
January	0	0
February	145	130.5
March	164	147.6
April	176	158.4
May	275	247.5
June	345	310.5
July	357	321.3
August	294	264.6
September	365	328.5
October	410	369.0
November	204	183.6
December	50	45.0

Table 4: Further estimate of potential quantities of lignocellulosic material.

Period	Days	Working days (Theoretical)	Theoretical production (t)	Assumed production (t)
March to August	184	127	1,587.5	1,587.5
September to February	181	124	1,550	1,162.5*
January to December	365	251	3,137.5	2,750
* Reduced value of 25% due to the climatic conditions characterizing this period (reduced service efficiency)				

Table 5: Amount of produced fresh and dried sludge.

Monthly sludge production 1	Monthly sludge production 2	Sludge average density 2	Monthly sludge production 2	Daily sludge production 2
(m^3/month)	(m^3/month)	(t/m^3)	(t/month)	(t/month)
27	13.5	0.694	9.4	0.310
1 = Fresh sludge leaving the anaerobic reactor; 2 = Dried sludge leaving the drying beds (arranged in cells)				

The sewage sludge (SS) production reported in Table 5, is very low due to the current operating conditions of the whole sanitation infrastructure. Data were given by ETAR. Once the availability of the different materials was known, using a modular approach, and establishing the initial input material as 1,500 t/y (125 t/month), it was possible to draw up a mixing plan on a monthly scale, that is reported in Table 6.

Table 6: Mixing strategy.

Raw material	Weight (t/month)	Weight (%)	A.D. (t/m^3)	Vol. (m^3)	Vol. (%)	SS (%)	SS (t)	SS (%)	N	C	C/N
OFMSW	78.1	62.5	0.75	104.1	49.3	25	19.5	44.8	1.4	30	21.4
Sludge	9.4	7.5	0.7	13.5	6.4	17	1.6	3.7	2.9	25	8.6
Ligno-cellulosic	37.5	30	0.4	93.8	44.3	60	22.5	51.5	0.7	45	64.3
Total	125	100	0.6	211.4	100	35	43.6	100	1.1	38	34.3

Table 7: Compost quantities produced.

COMPOST	Weight (t/y)	Volume (m^3/y)	Weight (t/d)	Volume (m^3/d)
Raw[1]	675*	1,269*	3.2	6
Fine[2]	450**	923**	2.1	4.4
1 = downstream of the maturation phase; 2 = downstream of the post-treatments				
* Raw Compost yield: 45% in Weight e 55% in Volume compared to the input quantity;				
** Finished Compost yield: 30% in Weight e 40% in Volume compared to the input				

The *first phase* of the process (accelerated bio-oxidation) is proposed to be carried in 3 cells and will be characterized by 21 days lasting (about 17–18 cycles/y). In each cell, about 29 tons of organic material by cycle can be treated. The aerobic conditions will be guaranteed by an insufflation system consisting of perforated corrugated tubes, a fan for the supply of air and one for suction. The exhausted process air will be sent to a bio-filter for its treatment before release into the atmosphere. This solution (used mainly from the mechanical biological treatment) is the most economic one and have a good pollutants removal efficiency [40]–[42]. Downstream of the accelerated bio-oxidation, a reduction in weight and in volume of 30% and 25% respectively, is was foreseen. In this way the quantity for the second phase is equal to 1,050 t/y (1,730.8 m^3/y).

The second phase is carried out by placing the material on a stave along the linear-shaped piles. This area will represent the largest area inside the station. The maturation stage will be characterized by a duration of about 60 days for a total number of 6 cycles/year, about 173 tons/cycle will be treated. It has been hypothesized 4 piles, with a base and length of 3 m and 32 m respectively (384 m^2 in plan covered with material in the secondary stabilization phase).

The piles are not aerated but they are subjected to turning operations thanks to a revolving machine, actuated and driven by the tractor, which plays a multifunctional role within the whole process. The wetting operations will be carried out by means of a rotary sprinkler located above the revolving machine. The overlapping of the two operations allows the optimization of the time-management aspect. In Table 7 the expected quantities of compost are reported.

For the OFMSW, a manual selection was considered for the removal of the bulkiest materials. For the brushwood, a bio-shredder was proposed. For the materials handling operations (mixing, heaps preparation and loading/unloading) a tractor with front blade can be used. At the end, a final screening trough a cylindrical roll was proposed. In this way, a homogenization of the material can be made but also will allow the recovery of any material not intercepted in the initial phases.

In Table 8 the surfaces needed for the proposed composting plant, considering the quantity of organic material that will be treated, are reported.

Given the above, the initial matrix value to be subjected to the treatment that was chosen to be considered is equal to 1,500 t/y. This assumption, following the natural reductions that occur during the proposed process, would allow a production of about 450–675 t/y, depending on whether the compost is considered refined or raw, or downstream or upstream of the post-treatments (dimensional screening), which allow a higher final quality.

Regarding the possible use of compost, the first and simplified land analysis has highlighted that its application can be suitable with the existing land and will improve its physical, chemical and biological characteristics. Its use in the agricultural field can be made only after the development of a proper campaign of dissemination, promotion and demonstration, in order that compost will be considered a product of value.

4 CONCLUSIONS AND PERSPECTIVES

The study carried out, has allowed highlighting how the considered city, Beira, presents a high and diversified potential of suitable available materials for the biodegradation process such as composting. The main contribution in the organic material productions is certainly due to market utilities; also, the catering activities represents an important contribution in terms of quality and diversification of the commercial activities involved.

With regard to the wood-cellulosic material, used as the bulking agent, there is a significant quantity, as the city has numerous tree-lined avenues. It must be considered that in Mozambique, wood is widely used in domestic combustion for cooking purposes. The sewage sludge, is taken in consideration with a view to an integrated approach of the utilities and materials involved. Its involvement covers the additive function, however if SS characteristics become unsuitable, it is possible to implement a mixing plan without this element.

Table 8: Internal surfaces of composting plant.

Areas		Surfaces (m^2)
Receiving and pre-treatments materials area	Green waste storage and shredding area	200
	Waste storage box	25
	Mixing box	25
Active composting area	3 Cells	75
	Biofilter	5
Maturation area	Heaps/windrow surface	390
	Lanes surface	260
	Operations surface	200
Post-treatments and storage compost area	Screening and storage compost area	240
Services area	Services local (/room)	40
Total		~1,500

Regarding the composting plant design, the choice to use a modular approach, i.e. to align the sizing to the quantity of input material established rather than the potentially available one, is configured as an evaluation tool for a successful project outcome. This, in fact, allows an analysis step by step of the real potential of using composting as an alternative to disposal in landfill, allowing a better understanding of the possible management limits and process difficulties on a small scale, minimizing the failure risk on a larger scale.

From a managerial point of view, for the success of the composting process, the logistic component must be taken into consideration by the personnel in charge. In this regard, the transmission and acquisition of adequate professional and managerial skills through a completed and iterated training is absolutely vital.

The composting process is capable of presenting its maximum effectiveness only within an ISWM. Composting is not and does not pretend to present itself as an absolute and exhaustive solution, but rather as an effective form of management of a given component of MSW, the biodegradable one.

In additional to the function of composting as a treatment form, the promotion of the "compost" use is important. This means that parallel projects that encourage the use of the product must be implemented. An example is the urban agriculture programs (agro-economy), which may prove capable of combining agricultural production activities for food purposes with the creation of useful spaces for the application of the obtained product, the compost (the country is currently undergoing strong timber exports).

Regarding the plant location, a crucial phase in the design of composting station, this remains a question not completely resolved within this study. Although some possible, suitable sites of the plant installation have been assessed, the selection of one rather than another must follows a careful and in-depth context analysis, also making use of geotechnical instruments. The choice, in addition to the various known criteria necessary to carry it out, should not be dictated by purely economic-financial issues in the short term. On the other hand, it cannot be exempt from careful assessments regarding medium/long-term, technical complications and burdens.

For the project implementation, at whatever scale it is carried out, a wide-ranging evaluation is absolutely necessary, which is capable of taking into consideration social, economic and environmental elements. It is essential, to not underestimate absolutely all those socio-educational and informative paths; that even if they seem apparently marginal, are fundamental ingredients for the success of the project, both from an acceptance and development point of view.

REFERENCES

[1] Serge Kubanza, N. & Simatele, M.D., Sustainable solid waste management in developing countries: a study of institutional strengthening for solid waste management in Johannesburg, South Africa. *Journal of Environmental Planning and Management*, **63**(2), pp. 175–188, 2020. https://doi.org/10.1080/09640568.2019.1576510.

[2] Ferronato, N., Torretta, V., Ragazzi, M. & Rada, E.C., Waste mismanagement in developing countries: A case study of environmental contamination. *UPB Scientific Bulletin*, **79**(3), pp. 185–196, 2017.

[3] Kiselev, A.V., Magaril, E.R. & Rada, E.C., Energy and sustainability assessment of municipal wastewater treatment under circular economy paradigm. *WIT Transactions on Ecology and the Environment*, vol. 237, WIT Press: Southampton and Boston, pp. 109–120, 2019. https://doi.org/10.2495/ESUS190101.

[4] Ragazzi, M., Rada, E.C., Torretta, V. & Salazar-Valenzuela, X., Management of urban wastewater on one of the Galapagos Islands. *Sustainability*, **8**(3), 208, 2016. https://doi.org/10.3390/su8030208.

[5] Nath, K.J. & Sengupta, A.K., An alternative approach for municipal wastewater management: Technology options for small and medium towns. *Water Practice & Technology*, **11**(1), pp. 157–165, 2016. https://doi.org/10.2166/wpt.2016.001.

[6] Gallego-Schmid, A. & Tarpani, R.R.Z., Life cycle assessment of wastewater treatment in developing countries: A review. *Water Research*, **153**, pp. 63–67, 2019. https://doi.org/10.1016/j.watres.2019.01.010.

[7] Omwoma, S., Lalah, J.O., Kueppers, S., Wang, Y., Lenoir, D. & Schramm, K.-W., Technological tools for sustainable development in developing countries: The example of Africa, a review. *Sustainable Chemistry and Pharmacy*, **6**, pp. 67–81, 2017. https://doi.org/10.1016/j.scp.2017.10.001.

[8] Senkoro, H., Solid waste in Africa: A WHO/AFRO perspective. CWG Workshop: Solid Waste Collection That Benefits the Urban Poor, Dar es Salaam. https://scialert.net/fulltext/?doi=jest.2001.93.102. Accessed on: 10 Feb. 2020.

[9] UNEP, 2002 International Source Book on Environmentally Sound Technology for Municipal Solid Waste Management. (IETC) Technical Publication, USA. https://scialert.net/fulltext/?doi=jest.2001.93.102. Accessed on: 10 Dec. 2020.

[10] Rai, R.K., Bhattarai, D. & Neupane, S., Designing solid waste collection strategy in small municipalities of developing countries using choice experiment. *Journal of Urban Management*, **8**(3), pp. 386–395, 2019. https://doi.org/10.1016/j.jum.2018.12.008.

[11] Ikhlayel, M., Development of management systems for sustainable municipal solid waste in developing countries: a systematic life cycle thinking approach. *Journal of Cleaner Production*, 180, pp. 571–586, 2018. https://doi.org/10.1016/j.jclepro.2018.01.057.

[12] Aleluia, J. & Ferrão, P., Assessing the costs of municipal solid waste treatment technologies in developing Asian countries. *Waste Management*, **69**, pp. 592–608, 2017. https://doi.org/10.1016/j.wasman.2017.08.047.

[13] Kerdsuwan, S., Laohalidanond, K. & Jangsawang, W., Sustainable development and eco-friendly waste disposal technology for the local community. *Energy Procedia*, **79**, pp. 119–124, 2015. https://doi.org/10.1016/j.egypro.2015.11.493.

[14] Ragazzi, M., Catellani, R., Rada, E.C., Torretta, V. & Salazar-Valenzuela, X., Management of municipal solid waste in one of the Galapagos islands. *Sustainability*, **6**(12), pp. 9080–9095, 2014. https://doi.org/10.3390/su6129080.

[15] Han, Z. et al., Influencing factors of domestic waste characteristics in rural areas of developing countries. *Waste Management,* **72**, pp. 45–54, 2018. https://doi.org/10.1016/j.wasman.2017.11.039.

[16] UNESCO and sustainable development. http://unesdoc.unesco.org/images/0013/001393/139369e.pdf. Accessed on: 10 Dec. 2020.

[17] Kattoua, M.G., Al-Khatib, I.A. & Kontogianni, S., Barriers on the propagation of household solid waste recycling practices in developing countries: State of Palestine example, *Journal of Material Cycles and Waste Management*, **21**(4), pp. 774–785, 2019. https://doi.org/10.1007/s10163-019-00833-5.

[18] Rada, E.C. & Cioca, L., Optimizing the methodology of characterization of municipal solid waste in eu under a circular economy perspective. *Energy Procedia*, **119**, pp. 72–85, 2017. https://doi.org/10.1016/j.egypro.2017.07.050.

[19] Sharma, G. et al., Gridded emissions of CO, NOx, SO2, CO2, NH3, HCl, CH4, PM2.5, PM10, BC, and NMVOC from open municipal waste burning in India. *Environmental Science and Technology*, **53**(9), pp. 4765–4774, 2019. https://doi.org/10.1021/acs.est.8b07076.

[20] Aleluia, J. & Ferrão, P., Characterization of urban waste management practices in developing Asian countries: A new analytical framework based on waste characteristics and urban dimension. *Waste Management*, **58**, pp. 415–429, 2016. https://doi.org/10.1016/j.wasman.2016.05.008.

[21] Ziraba, A.K., Haregu, T.N. & Mberu, B., A review and framework for understanding the potential impact of poor solid waste management on health in developing countries. *Archives of Public Health*, **74**(1), 55, 2016. https://doi.org/10.1186/s13690-016-0166-4.

[22] Dladla, I., Machete, F. & Shale, K., A review of factors associated with indiscriminate dumping of waste in eleven African countries. *African Journal of Science, Technology, Innovation and Development*, **8**(5–6), pp. 475–481, 2016. https://doi.org/10.1080/20421338.2016.1224613.

[23] Nagpure, A.S., Ramaswami, A. & Russell, A., Characterization the spatial and temporal patterns of open burning of municipal solid waste (MSW) in Indian cities. *Environmental Science and Technology*, **49**(21), pp. 12911–12912, 2015. https://doi.org/10.1021/acs.est.5b03243.

[24] Vithanage, M., Wijesekara, S.S.R.M.D.H.R., Siriwardana, A.R., Mayakaduwa, S.S. & Ok, Y.S., Management of municipal solid waste landfill leachate: A global environmental issue. *Environmental Deterioration and Human Health: Nat Anthropog Determ*, pp. 263–288, 2014. https://doi.org/10.1007/978-94-007-7890-0_11.

[25] Cocarta, D.M., Rada, E.C., Ragazzi, M., Badea, A. & Apostol, T., A contribution for a correct vision of health impact from municipal solid waste treatments. *Environmental Technology*, **30**(9), pp. 963–968, 2009. https://doi.org/10.1080/09593330902989958.

[26] Dos Muchangos, L.S., Tokai, A. & Hanashima, A., Application of material flow analysis to municipal solid waste in Maputo City, Mozambique. *Waste Management and Research*, **35**(3), pp. 253–266, 2017. https://doi.org/10.1177/0734242X16678067.

[27] Intharathirat, R., Abdul Salam, P., Kumar, S. & Untong, A., Forecasting of municipal solid waste quantity in a developing country using multivariate grey models. *Waste Management*, **39**, pp. 3–14, 2015. https://doi.org/10.1016/j.wasman.2015.01.026.

[28] Villalba, L., Donalisio, R.S., Cisneros Basualdo, N.E. & Noriega, R.B., Household solid waste characterization in Tandil (Argentina): Socioeconomic, institutional, temporal and cultural aspects influencing waste quantity and composition. *Resources, Conservation and Recycling*, **152**, 104530, 2020. https://doi.org/10.1016/j.resconrec.2019.104530.

[29] Nagpure, A.S., Assessment of quantity and composition of illegal dumped municipal solid waste (MSW) in Delhi. *Resources, Conservation and Recycling*, **141**, pp. 54–60, 2019. https://doi.org/10.1016/j.resconrec.2018.10.012.

[30] Ramesh, N., Meenambal, T., Vennila, G. & Ramesh, S., Start up, operations and monitoring of upflow anaerobic sludge blanket (UASB) reactor for the treatment of organic fraction of Municipal solid waste. *Ecology, Environment and Conservation*, **22**, pp. S319–S322, 2016.

[31] Ferronato, N., Preziosi, G., Gorritty Portillo, M.A., Guisbert Lizarazu, E.G. & Torretta, V., Assessment of municipal solid waste selective collection scenarios with geographic information systems in Bolivia. *Waste Management*, **102**, pp. 919–931, 2020. https://doi.org/10.1016/j.wasman.2019.12.010.

[32] Makarichi, L., Kan, R., Jutidamrongphan, W. & Techato, K.-A., Suitability of municipal solid waste in African cities for thermochemical waste-to-energy conversion: The case of Harare Metropolitan City, Zimbabwe. *Waste Management and Research*, **37**(1), pp. 83–94, 2019. https://doi.org/2019. 10.1177/0734242X18804029.

[33] Azevedo, B.D., Scavarda, L.F. & Caiado, R.G.G., Urban solid waste management in developing countries from the sustainable supply chain management perspective: A case study of Brazil's largest slum. *Journal of Cleaner Production*, **233**, pp. 1377–1386, 2019. https://doi.org/10.1016/j.jclepro.2019.06.162.

[34] Mohee, R., Mauthoor, S., Bundhoo, Z.M.A., Somaroo, G., Soobhany, N. & Gunasee, S., Current status of solid waste management in small island developing states: A review. *Waste Management*, **43**, pp. 539–549, 2015. https://doi.org/10.1016/j.wasman.2015.06.012.

[35] Brueckner, J.K., Slums in developing countries: New evidence for Indonesia. *Journal of Housing Economics,* **22**(4), pp. 278–290, 2013. https://doi.org/10.1016/j.jhe.2013.08.001.

[36] Waste Management Plan. Mozambique LNG. www.mzlng.com/content/documents/MZLNG/EIA/Volume_III/English/Annex_E_-_LNG_Final_EIA_Sept_2014_Eng.pdf. Accessed on: 10 Jan. 2020.

[37] Perteghella, A. & Vaccari, M., Organic waste valorization through composting process: A full-scale case study in maxixe, Mozambique. *Environmental Engineering and Management Journal*, **16**(8), pp. 1819–1826, 2017.

[38] CAM – Consorzio Associazioni con il Mozambico (Consortium of Associations with Mozambique), 2017. Campagna di Analisi Merceologica del Rifiuto Solido Urbano Domestico per la Città di Beira (Campaign of Commodity Analysis of Domestic Urban Solid Waste for the City of Beira). www.trentinomozambico.org/chi.html. Accessed on: 10 Dec. 2020.

[39] CAM – Consorzio Associazioni con il Mozambico (Consortium of Associations with Mozambique), 2017, Progetto di supporto tecnico al CMB nel settore dei RSU, Report sui Mercati cittadini che producono RSU – Organico (Project of technical support to the CMB in the sector of MSW, Report on City Market that certified MSW – Organic). www.trentinomozambico.org/chi.html. Accessed on: 10 Dec. 2020.

[40] Reyes, J. et al., Environmental performance of an industrial biofilter: Relationship between photochemical oxidation and odorous impacts. *Environmental Research*, **183**, 109168, 2020. https://doi.org/10.1016/j.envres.2020.109168.

[41] Schiavon, M., Ragazzi, M., Torretta, V. & Rada, E.C., Comparison between conventional biofilters and biotrickling filters applied to waste bio-drying in terms of atmospheric dispersion and air quality. *Environmental Technology*, **37**(8), pp. 975–982, 2016. https://doi.org/10.1080/09593330.2015.1095246.

[42] Cordova, T., Goodwin, J., Card, T. & Schmidt, C.E., Possitive air, biofilter cover layer control biosolids composting emission. *BioCycle*, **56**(3), pp. 49–51, 2015.

EFFECT OF SURCHARGE HEIGHT AND PRELOADING TIME ON LONG-TERM SETTLEMENT OF CLOSED LANDFILLS: A NUMERICAL ANALYSIS

MEHRNAZ ALIBEIKLOO[1], HAJAR SHARE ISFAHANI[2] & HADI KHABBAZ[3]
[1]Faculty of Engineering and IT, University of Technology, Australia
[2]Department of Civil Engineering, Isfahan University of Technology, Iran
[3]Faculty of Engineering and IT, University of Technology, Australia

ABSTRACT

In recent years, by developing cities and increasing population, reconstructing on closed landfill sites is unavoidable in some regions. Long-term settlement is one of the major concerns associated with reconstruction on landfills after closure. The purpose of this research is evaluating the effect of preloading in various patterns of height and time on long-term settlements of closed landfills. In this regard, five scenarios of surcharge from 1 to 3 m high within 3, 4.5 and 6 months of preloading time have been modelled using PLAXIS 2D software. Moreover, the numerical results have been compared to those obtained from analytical methods, and a good agreement has been achieved. The findings indicate that there is a linear relationship between settlement and surcharge height. Although, long-term settlement decreased by applying a longer and higher preloading, the time of preloading was found to be a more effective factor compared to preloading height.

Keywords: preloading, long-term settlement, closed landfill, PLAXIS 2D.

1 INTRODUCTION

Nowadays, considering the population growth, the cities are developing, which can lead to place some closed landfill sites within urban areas. Therefore, in some cases reconstruction on closed landfill is inevitable. Roads and public transport network are most probable cases for reconstruction on the top of closed landfill sites. There are a variety of challenges related to redevelopment on the top of closed landfills including, long term settlement control and protecting the stability of landfill structure. In order to reduce the super-structure settlement, improving the settlement characteristics of waste material is necessary. Moreover, the used method for improving the mechanical properties of waste disposal should be safe for the landfill cover layer and the drainage system.

Different methods can be applied for improving the ground, including stone columns, cement grouting, chemical stabilization, dynamic compaction and preloading [1]. However, some of ground improvement methods are not appropriate for landfills as a result of heterogeneous nature of wastes and existing different landfill components such as the liner structure. As an example, intrusive improvement techniques such as dynamic compaction (DC) can destroy the landfill cover layer, which usually consist of compacted clay with a synthetic barrier or geosynthetic clay liner (GCL) as a hydraulic barrier. Among the existing improvement techniques, preloading is a safe and non-intrusive method. Preloading is applying an external load on a low-permeable soil in order to accelerate the consolidation process and increasing the effective stress [2]. Referring to [3], the process of preloading is a simple form of applying induced settlement. Since wastes in landfills have heterogeneous nature, predicting settlement, especially secondary settlement is a major challenge. Therefore, there are limited studies associated with applying preloading technique on top of landfills. The most recent case study of using preloading for landfill improvement have been presented by [4] at Fresh Kills landfill on Station Island, New York. The landfill classification was the mixture of municipal solid wastes and construction debris. According to the results

WIT Transactions on Ecology and the Environment, Vol 247, © 2020 WIT Press
www.witpress.com, ISSN 1743-3541 (on-line)
doi:10.2495/WM200081

of this research, it was concluded that the predicted settlement under the embankment was from 0.15 m to 1.83 m. The initial measurements indicated that about 0.5 m of the predicted settlement occurred during the first 1.5 months. Moreover, the settlement parameters were calculated for wastes at Fresh Kills landfill using back analysis method, which are as follows [4]:

- The average value for modified secondary compression indices is 0.022.
- The average values for modified primary and recompression indexes are 0.23 and 0.023, respectively.

There is another case study at Tinton Falls landfill in New Jersey, USA [5]. In this site, both method of dynamic compaction and preloading were used aiming to improve landfill behaviour. The results of this case study illustrated that the secondary compression index was between 0.059 and 0.0817, which was considerably different with the values presented by [4] and showed the variation of settlement parameters for different wastes in various landfill classifications. Furthermore, it is outlined that observed total and differential settlements for preloading zone was higher than those observed in the zones where dynamic compaction were applied [5].

Aforementioned, preloading is a very simple (no need to employ special equipment), non-intrusive and an effective method (even for deeper improvements); however, it is a time consuming method. Although using a vertical drainage system is a common method to decrease the time of applying surcharge, vertical drains cannot be used in the landfill because they may destroy the landfill cover layer. Vertical drains can also provide a route in which the hazardous substance can be released [6]. In order to make preloading a time effective method, increasing the height of surcharge can be proposed as a suitable alternative. Having the knowledge of long-term settlement of landfills based on time and height of preloading leads to optimize the preloading design. As a result, in this research the effect of preloading on long-term settlements in closed landfill sites, considering the time and height of surcharge, has been investigated using numerical and theoretical methods.

2 THEORY OF SETTLEMENT

There are a variety of factors contributing to landfill settlements including, type of waste, organic content, moisture content, compaction density, compressibility, and level of nutrients available for biological activities, pH, temperature, and time [7]. Landfill settlement is considerably complex in comparison with coarse and fine soils as a result of heterogeneous nature of wastes, large particle sizes, compression of refuse particles, and the loss of solid due to biodegradation [8]. In addition, landfill settlements are classified into five groups [9]:

1. Initial compression: This settlement is the immediate mechanical compression caused by compression of highly deformable waste components and takes place immediately after an external load applied to the waste.
2. Primary settlement: It is the mechanical settlement due to continuous compression or reorientation of waste which is a slower process. This settlement is the consolidation settlement that occurs due to dissipation of water and gas from the pores as a result of applying the load.
3. Secondary settlement: This settlement is the mechanical compression due to the creep and primary decomposition of the waste.
4. Decomposition settlement: The settlement as a result of decomposition of organic material.
5. Residual settlement: Mechanical deformation and organic decomposition cause residual settlement.

Fig. 1 presents the settlement stages of wastes based on categories mentioned above. According to this figure, the settlement due to the self-weight or surcharge takes place within 3 months. The settlement at stages III to V is highly substantial and is time-dependent. Decomposition settlement is considerably important at landfills with organic and putrescible materials. In this research, it has been assumed that the waste materials are old and mostly inorganic and non-putrescible. Therefore, decomposition settlement due to biodegradation of putrescible material is disregarded.

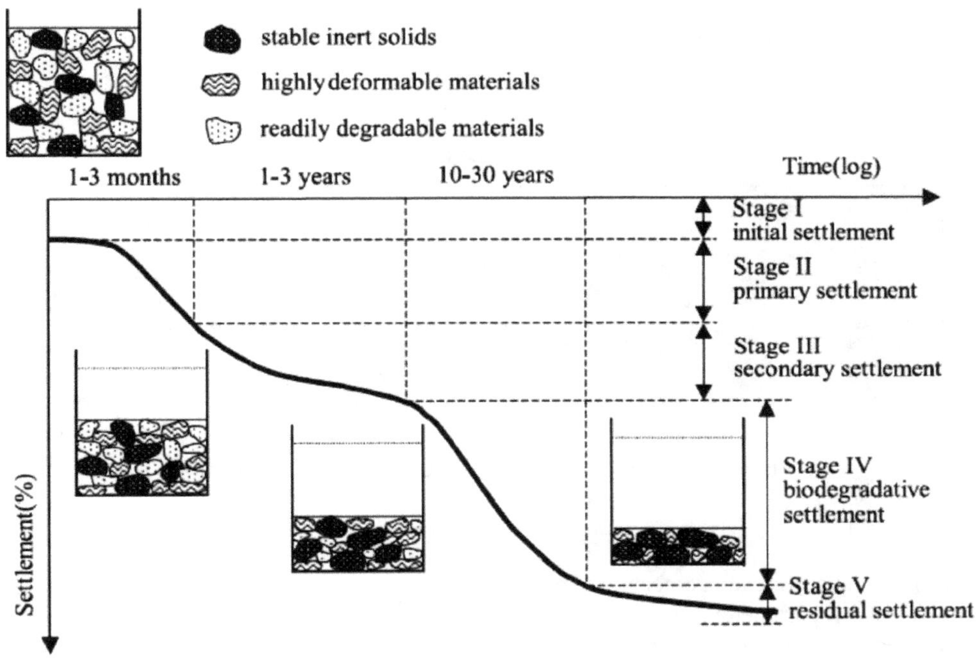

Figure 1: The typical time-settlement data at landfills [9].

There are many investigations associated with the theory and the methods to evaluate the primary and long-term settlements at landfills [10]–[12]. The following equations for estimation the primary (ΔH_P) and secondary settlement (ΔH_s) has been presented [11]

$$\Delta H_P = HC_{ce}\log\frac{P_0+\Delta P}{P_0} \quad \text{where } C_{ce} = \frac{C_c}{1+e_0}, \tag{1}$$

$$\Delta H_s = H_1 C_{\alpha\varepsilon}\log\frac{t_2}{t_1} \quad \text{where } C_{\alpha\varepsilon} = \frac{C_\alpha}{1+e_p}, \tag{2}$$

where, H is the height of waste layer, H_1 is the height of waste layer at the beginning of the secondary settlement, C_c is the primary compression index, e_0 is the initial voids ratio of waste, P_0 is the existing pressure at midlevel of layer, ΔP is the increase in overburden pressure acting at midlevel of layer, C_α is the secondary compression index, e_p is the void ratio after primary compression, t_2 is the time, t_1 is the time which is necessary for primary compression.

3 MATERIALS AND METHODS

The geometry of landfill considered for evaluating the effect of preloading is shown in Fig. 2. According to this section, total height of landfill is 20 m with 5 m above the natural ground level and the thickness of cover layer is assumed 0.5 m. As mentioned earlier, in this research, the waste material has been assumed non-putrescible with no biodegradation settlement. In order to calculate the long-term settlement, a detailed numerical analysis using PLAXIS 2D has been carried out, and the results have been verified with those obtained by analytical methods.

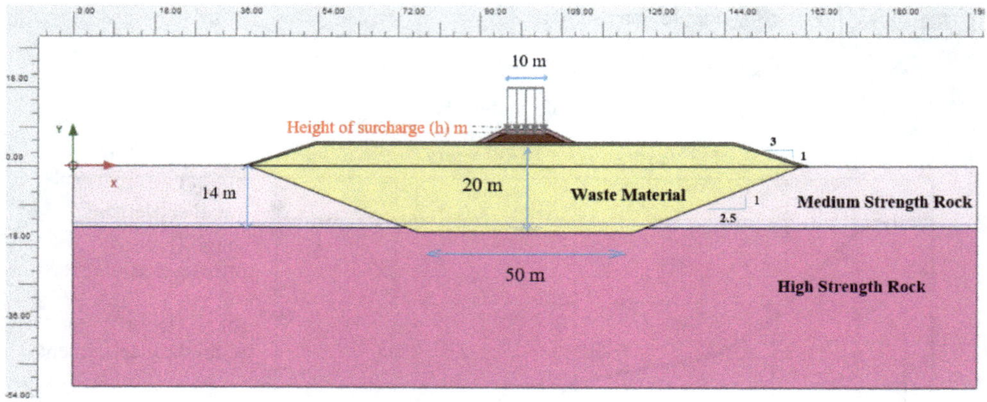

Figure 2: A typical cross section of the proposed landfill.

3.1 Material

In order to determine the geotechnical parameters of materials, the most effective parameters are settlement parameters of waste materials, including the primary compression index C_{ce} and the secondary compression index $C_{\alpha\varepsilon}$. There are some references reporting different values for C_{ce} and $C_{\alpha\varepsilon}$ in waste materials. According to Table 1, [13] reported the values of $C_{\alpha\varepsilon}$ for various landfill ages. It can be concluded that values of $C_{\alpha\varepsilon}$ ranges from 0.001 to 0.24 and it is difficult to use one $C_{\alpha\varepsilon}$ for the entire period of landfill. Moreover, there is another literature review conducted by [14]; and data were collected from several published studies. The most recent and relevant data of these investigations are illustrated in Table 2.

Based on the value of settlement parameters, mentioned in the literature, and based on the assumed condition for the waste material (non-putrescible), the geotechnical and design parameters considered for landfill and sub-soil layers are summarized in Table 3.

Table 1: Secondary compression parameters for waste material [13].

Material	$C_{\alpha\varepsilon}$
Ten year old landfill	0.02
Fifteen year old landfill	0.24
Fifteen to twenty year old landfill	0.02
Old landfill	0.04
Old landfill with high soil content	0.001 to 0.005

Table 2: Summary of C_{ce} and $C_{\alpha\varepsilon}$ parameters from literature [14].

Primary C_{ce}	Secondary $C_{\alpha\varepsilon}$	Reference
0.09–0.19	0.006–0.012	[15]
0.16	0.02	[16]
0.17–0.24	0.01–0.016	[17]
0.073–0.132	0.015–0.03	Field monitoring, [18]
0.17–0.23	0.024–0.030	Field monitoring, [19]
–	$C_{\alpha\varepsilon(EL)}$:0.02[a] $C_{\alpha\varepsilon(SW)}$: 0.19–0.28[b]	[20]
–	$C_{\alpha\varepsilon(EL)}$: 0.014 for waste treated with DDC[c] $C_{\alpha\varepsilon(EL)}$: 0.045 for waste treated with surcharge	Field monitoring for external load [5]

[a]EL = External load; [b]SW = Self-weight; [c]DDC = Deep dynamic compaction.

Table 3: Proposed geotechnical and design parameters for assumed landfill and sub-soil layer.

Material type	Unit weight (kN/m³)	Effective cohesion (kPa)	Effective friction angle (degree)	$C_{\alpha\varepsilon}$	C_{ce}
Medium strength rock	21	10	30	–	–
High strength rock	21	200	30	–	–
Waste	15	0	30	0.02	0.15
Cover layer	20	10	25	–	–
Embankment	21	2	33	–	–
Upper fill (extra surcharge)	21	10	30	–	–

3.2 Methods

In this research, the numerical simulation is performed using finite element analysis software PLXIS-2D which is a powerful software for two dimensional analysis of deformation in geotechnical engineering. The finite element mesh of the landfill section with 15 node triangular element is shown in Fig. 3.

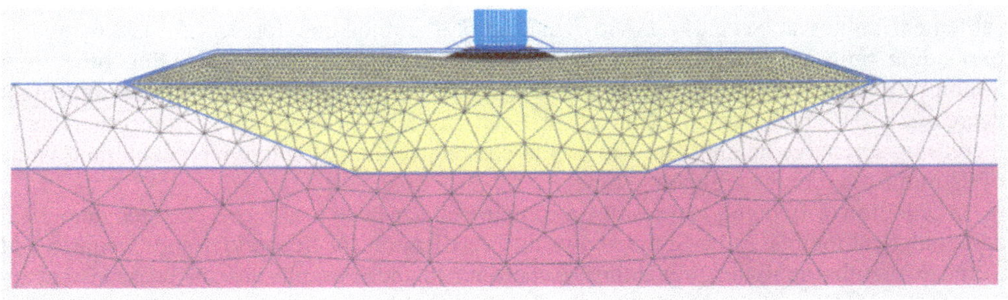

Figure 3: Deformed mesh of the model.

The soft soil creep (SSC) model has been considered for calculation of creep (secondary settlement). This model is very common for soft soils such as normally consolidated clays, peats and wastes and this model in PLAXIS computes the secondary compression. Since soft soils are highly compressive, in the oedometer test normally consolidated clay behaves ten times softer than normally consolidated sand [21]. In addition, oedometer stiffness of soft soil has linear dependency and in the stress-stiffness curve a line in the form $E_{oed} = \sigma/\lambda^*$ can be plotted, where λ^* is the modified compression index. The SSC model is an extension of soft soil (SS) model, which is based on Cam Clay model considering Mohr Coulomb criterion for failure. In soft soil (SS) model, increasing the load would cause the stress state to fall outside the current cap and the cap expands instantaneously, while in the SSC model this cap shift needs time. It means that by applying higher load, the cap does not expand immediately and it may take one day to adopt to the new stress state [22]. In order to calculate creep settlement, [23] proposed the following equation:

$$\varepsilon^H = \varepsilon_C^H + \mu^*.\,ln\left(\frac{\tau_c+t'}{\tau_c}\right), \tag{3}$$

where, ε_C^H is consolidation settlement, μ^* denotes the creep index describing secondary compression per logarithmic time increment, and τ_c and t' are time parameters. The total volumetric strain in an isotropic stress state as a result of increasing the mean effective stress from p'_0 to p' during the time of $(\tau_c + t')$ is expressed as:

$$\varepsilon_v = \kappa^*.\,ln\left(\frac{p'}{p'_0}\right) - (\lambda^* - \kappa^*).\,ln\left(\frac{p'_{pc}}{p'_{p0}}\right) + \mu^*.\,ln\left(\frac{\tau_c+t'}{\tau_c}\right). \tag{4}$$

In eqn (4), the parameters p'_{p_0} and p'_{pc} represent the pre-consolidation pressure relating to before loading and end of consolidation, respectively and κ^* is the modified swelling, which determines soil behaviour during unloading/reloading. The relationships between basic stiffness $(\kappa^*, \lambda^*, \mu^*)$ an internationally normalized parameters (C_r, C_c, C_α) are [24]:

$$\lambda^* = \frac{C_c}{2.3(1+e)}, \kappa^* = \frac{2}{2.3}\frac{C_r}{1+e}\ \mu^* = \frac{C_\alpha}{2.3(1+e)}. \tag{5}$$

4 RESULTS AND DISCUSSION

The obtained numerical results for the five evaluated scenarios at three different preloading time are presented in Table 4. In this table, h_s is the surcharge height, which is assumed ranging from 1 m to 3 m with 0.5 m increment in each scenario and t_p is the preloading time, which is considered to be 3, 4.5, and 6 months. It is obvious that after removing surcharge and applying the reconstruction load, the value of settlement increases with time and this settlement is calculated up to 10 years of applying external load. As an example, the settlement values at Scenario 1 and Scenario 5 of preloading for 3, 4.5, and 6 months preloading time are plotted in Fig. 4. As expected, by increasing preloading time from 3 months to 6 months, the long-term settlement decreased. The maximum of this settlement decrease is 319 mm, which occurred at Scenario 5 of preloading. On the other hand, increasing surcharge height from 1 m to 3 m can reduce the settlement with maximum value of 83 mm for 6 months preloading time.

In order to verify numerical results, a comparison has been performed between numerical and analytical method [11]. Fig. 5 shows this verification for Scenario 1 and Scenario 3 at 3 and 6 months applying preloading. According to these graphs, it can be seen that by increasing the time of applying the reconstruction load, the difference between calculated settlement using theoretical and numerical methods are developing. However, the maximum

Table 4: Numerical results for the five evaluated scenarios at different preloading time.

Scenario	t_p months	Settlement (mm)										
		Different times of applying reconstruction load (year)										
		0.5	**1**	**2**	**3**	**4**	**5**	**6**	**7**	**8**	**9**	**10**
1, hs = 1 m	**3**	508	675	840	934	999	1,049	1,089	1,123	1,152	1,177	1,201
	4.5	375	522	675	764	828	876	916	950	979	1,004	1,027
	6	292	423	566	652	714	762	801	834	863	888	911
2, hs = 1.5 m	**3**	475	644	809	905	972	1,023	1,064	1,099	1,129	1,156	1,180
	4.5	344	491	647	738	803	853	894	928	958	984	1,007
	6	256	387	532	620	682	731	771	805	833	859	882
3, hs = 2 m	**3**	454	623	795	885	961	1,014	1,056	1,092	1,123	1,150	1,175
	4.5	317	463	620	715	778	831	872	908	939	965	989
	6	229	361	508	596	660	710	750	785	815	841	865
4, hs = 2.5 m	**3**	429	604	775	874	944	997	1,040	1,076	1,106	1,135	1,160
	4.5	290	437	598	692	759	810	852	888	918	945	970
	6	208	336	485	574	639	689	730	765	795	821	845
5, hs = 3 m	**3**	408	579	755	856	926	981	1,024	1,061	1,094	1,122	1,147
	4.5	269	414	575	670	738	791	834	870	901	927	952
	6	183	316	463	554	619	669	710	745	776	803	828

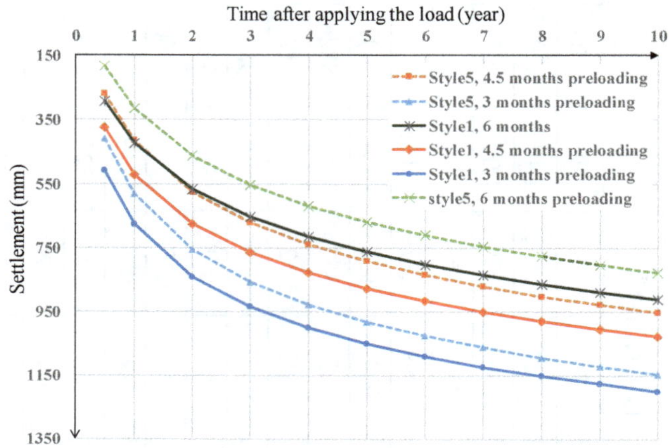

Figure 4: Settlement versus time after applying the reconstruction load at Scenario 1 and Scenario 5 preloading.

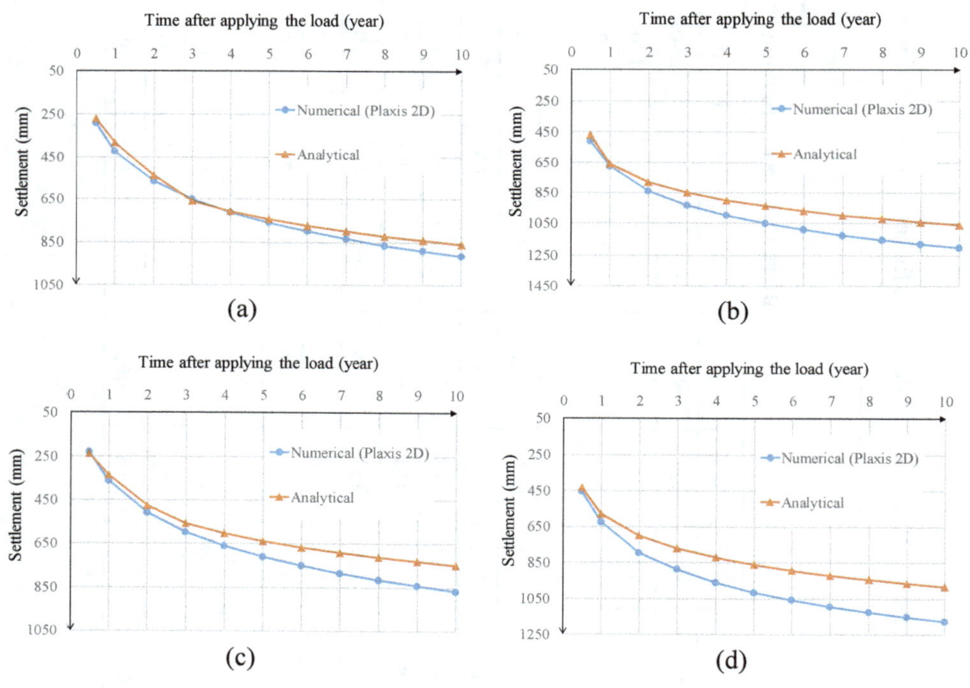

Figure 5: Verification of numerical results with analytical results. (a) Scenario 1, 6 months; (b) Scenario 1, 3 months; (c) Scenario 3, 6 months; and (d) Scenario 3, 3 months.

value of this difference which is related to Scenario 3 and 3 months preloading, is limited to 17% and this value of disagreement is reasonable. Therefore, the numerical and analytical results are in acceptable agreement with each other.

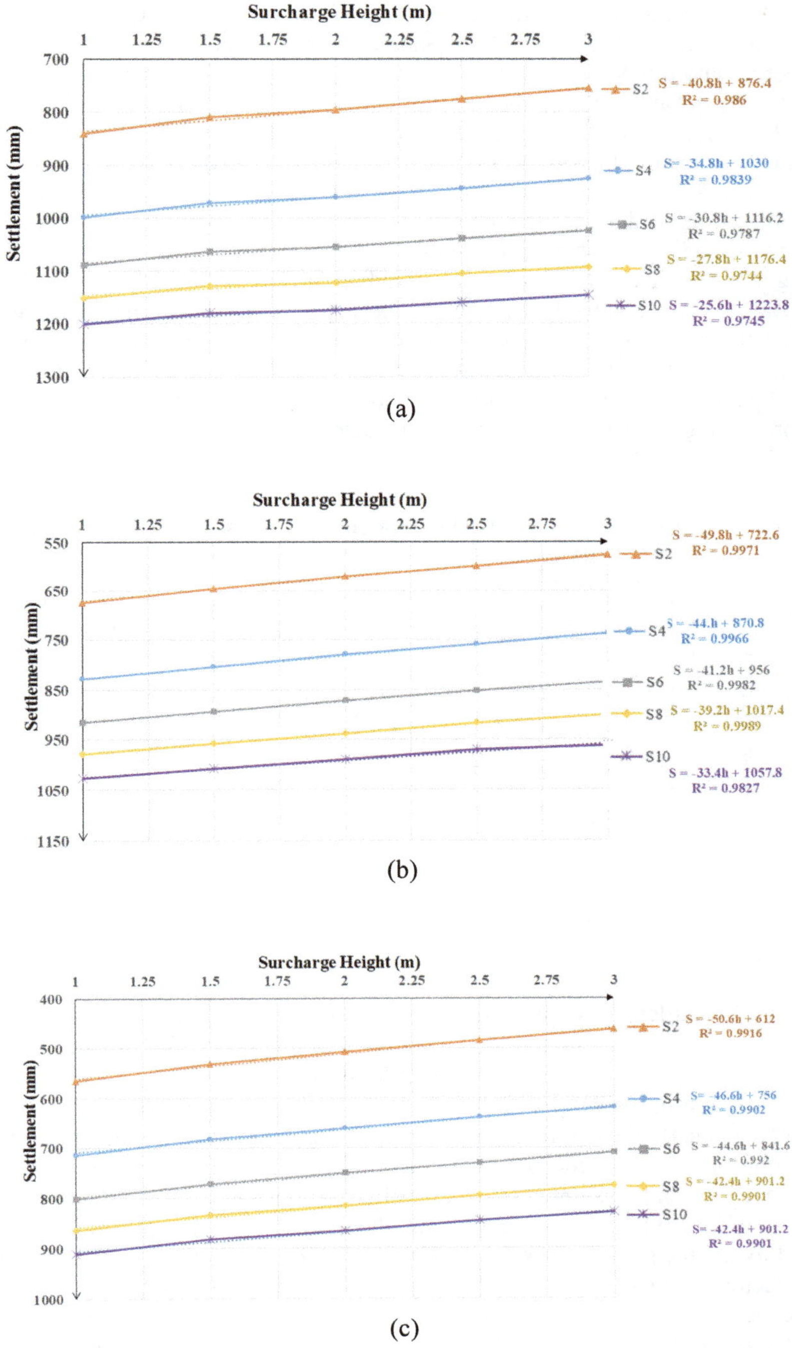

Figure 6: Settlement versus surcharge height for 2, 4, 6, 8 and 10 years after applying the reconstruction load. (a) 3 months; (b) 4.5 months; and (c) 6 months.

In Fig. 6, the variation of settlements versus surcharge height after different times of applying reconstruction load (2, 4, 6, 8, and 10 years) are presented. According to these graphs, there is approximately a linear relationship between the long-term settlement and the height of surcharge, and the corresponding equation is presented close to each graph. On the other hand, the rate of settlement reduction decreases by increasing the time of applying external load, indicating that in the first years of applying reconstruction loads, increasing the surcharge height has more effect on decreasing settlements. In order to evaluate the effect of preloading time and surcharge height in reducing the settlement of landfills under reconstruction loads, the settlement values after 10 years of applying external loads and preloading time for different scenarios of preloading (different surcharge heights) are plotted in Fig. 7. Based on the obtained results, by increasing the preloading time for 1.5 months, the long-term settlement decreases about 15% while by increasing the surcharge height from 1 m to 3 m, the settlement values decrease approximately 5%. As a result, by comparing these two crucial parameters in the preloading ground improvement method, it can be concluded that the preloading time is a more important factor compared to the surcharge height in curtailing the settlement.

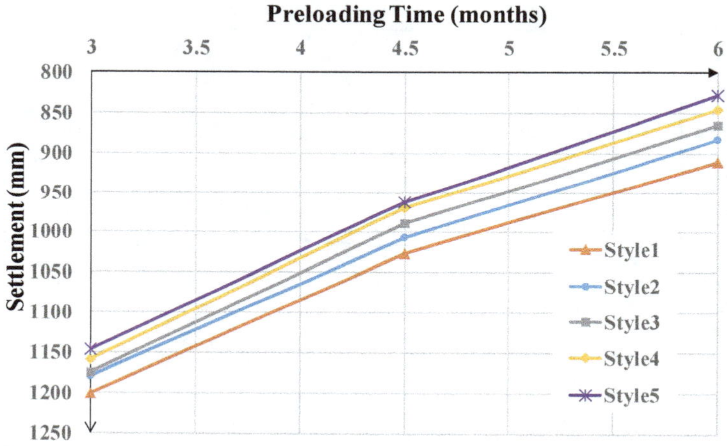

Figure 7: Settlement versus preloading time for different surcharge scenarios.

5 CONCLUSIONS

Preloading is a non-intrusive ground improvement technique at closed landfill sites, which can prevent damages to the landfill structures, such as drainage systems and liner cover, under reconstruction loads. In this method, height and time of applying surcharge are two crucial factors to save time and make projects be cost effective. In other words, by selecting a proper combination of height and time of preloading, the time of ground improvement method can be reduced, while the project costs can be kept under control. In this research, the effects of these two factors on long-term settlement of waste material deposits have been investigated by conducting an array of numerical analyses with PLAXIS 2D. The predicted results were verified with an existing theory of settlement. In the assumed conditions, the results indicated that, preloading time was a more effective parameter compared to the surcharge height. Moreover, the settlement values for different surcharge height varied linearly; and this linear relationship was developed for each fitted line.

To conclude, in projects in which preloading is a proposed method of ground improvement, a reasonable combination of preloading height and time can be opted by considering the maximum allowable settlement. There might be some time, cost, and construction limitations in some projects, which can affect the optimum combination of preloading time and height. Hence, waste materials can be managed for reconstruction purposes by adopting preloading method as a non-intrusive method to control long-term settlement of landfills. Since preloading approach does not disturb the internal systems of closed landfills (e.g. cover, drainage, liner, gas collection and leachate collection systems), it can be considered for redevelopment in areas with land scarcity employing environmental concerns.

REFERENCES

[1] Bouazza, A. & Van Impe, W.F., Ground modification methods applied to municipal solid waste landfills. *Australian Geomechanics*, pp. 73–83, 1997.

[2] Moseley, M.P. & Kirsch, K., *Ground Improvement*, 2nd ed., Taylor and Francis, 2004.

[3] Mohamad, R., Precompression of soft soils by surcharge preloading: Some common pitfalls and misunderstood fundamentals, 2008.

[4] De Melo, L., El-Sherbiny, R., Steier, W. & Salem, M., Evaluation of waste compressibility due to preloading at the Fresh Kills Landfill. *GeoCongress 2012: State of the Art and Practice in Geotechnical Engineering*, pp. 4184–4193, 2012.

[5] Lewis, P.J., Mansfield, J., Ashraf, S. & Zicko, K., Performance of a highway embankment constructed over landfill material. *Proceedings of the Fifth International Conference on Case Histories in Geotechnical Engineering*, New York, 2004.

[6] Mitchell, J.K. & Van Court, W.A., Role of soil modification in environmental engineering applications. *Geotechnical Special Publication*, **1**(30), pp. 110–143, 1992.

[7] Tahmoorian, F., Development of a technical management tool for settlement and stability behaviour of municipal solid waste landfills. Master's thesis, University of Technology Sydney, Australia, 2014.

[8] Emberton, J. & Parker, A., The problems associated with building on landfill sites. *Waste Management and Research*, **5**(1), pp. 473–482, 1987.

[9] Grisolia, M., Napoleoni, Q. & Tangredi, G., The use of triaxial tests for the mechanical characterization of municipal solid waste. *Proceedings of the Sardinia 1995, Fifth International Landfill Symposium*, **2**, pp. 761–767, 1995.

[10] Ling, H.I., Leshchinsky, D., Mohri, Y. & Kawabata, T., Estimation of municipal solid waste landfill settlement. *Journal of Geotechnical and Geoenvironmental Engineering*, **124**(1), pp. 21–28, 1998.

[11] Sowers, G.F., Settlement of waste disposal fills. *Proceedings of the 8th International Conference on Soil Mechanics and Foundation Engineering*, Moscow, pp. 207–210, 1973.

[12] Dodt, M.E., Sweatman, M.B. & Bergstorm, W.R., Field measurements of landfill surface settlements. *Geotechnical Practice for Waste Disposal, ASCE*, Reston, VA, **7**, pp. 406–417, 1987.

[13] Oweis, I.S. & Khera, R., *Geotechnology of Waste Management*, 2nd ed., PWS Publishing Company, 1998.

[14] Sharma, H.D. & De, A., Municipal solid waste landfill settlement: Postclosure perspectives. *Journal of Geotechnical and Geoenvironmental Engineering*, **133**(6), pp. 619–629, 2007.

[15] Boutwell, G.P. & Fiore, V.A., Settlement of clay cover on saturated garbage. *Geoenvironment 2000: Characterization, Containment, Remediation, and Performance in Environmental Geotechnics*, ASCE, pp. 964–979, 1995.

[16] Stulgis, R.P., Soydemir, C. & Telgener, R.J., Predicting landfill settlement. *Geoenvironment 2000, Geotechnical Special Publication ASCE*, **2**(1), pp. 980–994, 1995.

[17] Landva, A.O., Valsangkar, A.J. & Pelkey, S.G., Lateral earth pressure at rest and compressibility of municipal solid waste. *Canadian Geotechnical Journal*, **37**(6), pp. 1157–1165, 2000.

[18] Marques, A.C.M., Filz, G.M. & Vilar, O.M., Composite compressibility model for municipal solid waste. *Journal of Geotechnical and Geoenvironmental Engineering*, **129**(4), pp. 372–378, 2003.

[19] Andersen, E.O., Balanko, L.A., Lem, J.M. & Davis, D.H., Field monitoring of the compressibility of municipal solid waste and soft alluvium. *International Conferences of Case Histories in Geotechnical Engineering*, 2004.

[20] Sharma, H.D., Solid waste landfills: Settlements and post-closure perspectives. *Environmental and Pipeline Engineering*, pp. 447–455, 2000.

[21] Gündüz, B., Analysis of settlements of test embankments during 50 years: A comparison between field measurements and numerical analysis. Master's thesis, Lund University, 2008.

[22] Waterman, D. & Broere, W., Practical application of the soft soil creep model. *Plaxis Bulletin*, **15**, 2004.

[23] Butterfield, R., A natural compression law for soils. *Géotechnique*, **29**(4), pp. 469–480, 1979.

[24] Neher, H., Wehnert, M. & Bonnier, P., An evaluation of soft soil models based on trial embankments. *Computer Methods and Advances in Geomechanics*, ed. C. Desai, AA Balkema: Rotterdam, pp. 373–378, 2000.

COMPOUNDING AND PROCESSING HYDRO-BIODEGRADABLE PLASTIC FILMS FOR PLASTIC WASTE REDUCTION. PART I: PROCESSING CONDITIONS AND ENVIRONMENTAL PERFORMANCE AGAINST PLASTIC SOLID WASTE

HAJAR JAWAD KARAM, MASUMAH AL-QASSIMI, MAJED HAMEED AL-WADI,
AISHA ABDULLAH AL-ROWAIH, ABDUL SALAM AL-HAZZA'A & SULTAN MAJED AL-SALEM
Environment & Life Sciences Research Centre (ELSRC), Kuwait Institute for Scientific Research (KISR), Kuwait

ABSTRACT

The reliance on conventional polymeric materials derived from petrochemical sources, has been associated with various environmental burdens and stressors that are well documented. These are namely associated with the overall carbon print of plastics conversion and an overwhelming accumulation of solid waste; which on the other hand is associated with soil and water aquifers poisoning in many regions the world over. In this work; and in an effort to develop a polymeric blend that can sustain itself in terms of mechanical integrity, thermoplastic starch polyester was converted with virgin linear low density polyethylene (LLDPE) and a plastic solid waste (PSW) component retrieved from municipal sources. The objective in this communication is to study the effect of different compounding conditions on the resulting blends which were characterized using scanning electron microscopy (SEM), mechanical profiling in accordance with ASTM 882-12 and physical-optical properties determination. The blends were originally prepared using single screw extrusion (SSE), twin screw extrusion (TSE) and blown-filming, to determine which of the processes are more reliable for industrial bulk production purposes in the future to establish a plastic conversion line suitable for biodegradable plastic film conversion.

Keywords: biodegradation, polyester, starch, polymer waste, waste management.

1 INTRODUCTION

Due to the environmental concerns associated with the disposal of plastic solid waste (PSW), the development of a new class of materials becomes a necessity rather than a luxury. Biodegradable plastics presents itself as such, and it has proven itself as an effective measure against conventional plastics. Comprehensive assessments of biodegradable polymers applications and current developments were reviewed previously by Amass et al. [1]. Emadian et al. [2] reported that biodegradation cannot lead to a complete disappearance of the material. The colonisation of microorganism on the plastic occurs when materials are exposed to the appropriate conditions of soil and humidity. Chiellini et al. [3] has also provided an overview of the degradable polymers in agricultural applications. Orhan and Büyükgüngör [4] tested the biodegradability of polyethylene (PE) bags containing 12% by weight of starch. They reported that PE plastics build-up in the environment is at a rate of 25 million tons per year. Ohtaki and Nakasaki [5] examined the degradability of several biodegradable plastics under controlled composting conditions and concluded that the degradability of 8 plastic ranged from a few percentage points to approximately 65% over the 8 days composting period at 50°C. Studies of polymer degradation during their lifecycle have shown that there are usually two main steps involved: abiotic and biotic degradation. Abiotic degradation involves the degradation of the polymeric content due to environmental exposure and is accompanied by microbial colonisation (biotic degradation) [6].

WIT Transactions on Ecology and the Environment, Vol 247,
www.witpress.com, ISSN 1743-3541 (on-line)
doi:10.2495/WM200091

Biodegradation at chemical level is initiated by containing a carbonyl functional group (C=O) within a chemical chain. Hydro-biodegradable materials contain carbonyl groups, which break down with heat and moisture. A 100% of such material is therefore classified on its own as biodegradable. Oxo-biodegradable plastics, do not contain carbonyl groups making it essential to introduce other additives which will change their chemical properties and accelerate the oxidation process leading to biodegradation of the material. Reddy et al. [7] revealed that biodegradables can mitigate harmful pollutants from leaching into surrounding environments, such as chlorinated compounds and toxins. The reader is referred to our earlier work on the biodegradability of oxo-biodegradables in arid climates as well [8]–[10].

Montagna et al. [11] also reported the use of different pro-degradant materials to enhance polymers biodegradability and showed the influence of an organic pro-degradant on PP under the two different environmental conditions (abiotic and biotic). According to literature, various characterisation methods such as spectroscopy (FTIR), mechanical testing [12], [13], and thermal analysis (TGA and DSC) [14] have been used to assess the composition of waste [15] as a result of their susceptibility to changes in material properties. FTIR spectroscopy is found to be the most effective and sensible technique for measuring polymer degradation [15]. In this article which is the first part of a two series communication, we present the compounding and basic characterisation of various biodegradable blends originating from virgin linear low density polyethylene (LLDPE) and PSW, enhanced with a prod-degradant and a thermoplastic starch (TPS) resin.

2 EXPERIMENTAL

Commercial grade LLDPE (EFDC-7050) was used as received in a pellet form with a reported density of 0.918 g cm^{-3} [8]. The LLDPE polymer was compounded with a 3 wt.% ECOVIO® (TPS) resin and by the same amount with a PDQ-M Willow Ridge Plastics (Inc.) pro-degradant chemical, using three different techniques that are standardised in the industry of plastics conversion. These techniques are namely, (a) single screw extrusion (SSE), (b) twin screw extrusion (TSE), and (c) blow-filming (BF) which are the most common among commercial film production lines, and have been chosen to study the effect of processing and melting on the materials. SSE was performed using a Dolci extruder (L/D ratio 1/28) using 10 kg as master batch mix. The blends with TPS were conducted in the same mix ratios as the oxo-additives. Mixing with palletized PSW was also conducted in the same blend ration of 3 wt.%. Furthermore, BF was conducted with the same formulations aforementioned using a Windsor blown-film coupled with an L/D (1/28) SSE operated under 6 bars of pressure. TSE was performed using a Kung Hsing operated in co-rotation with a maximum speed of 75 RPM maintained throughout the experimental work. The TPS samples were extruded in a standalone Brabender SSE which was used in the compounding process of the samples. All samples were extruded as sheets of 100 μm thickness meeting standard market specifications and as an average of major film applications. Mechanical properties were studied following test protocol of ASTM 882 using a Tinius Olsen Universal Testing Machine (UTM) machine as depicted previously in Al-Salem et al. [8]. The optical and physical properties of the samples were evaluated using the PCE-CSM3 colorimeter following ASTM D 2244 protocol to measure colour parameters. The tests were used to establish the colour parameters for light/dark (L^*), red/green (a^*), and yellow/blue (b^*). The average readings of the specimen are reported. The haze (%) and light transmission (%) of the samples were determined using a MF709 (spherical) haze meter (model BS 2782) following ASTM D 1003 similar the methods applied in past works [8], [16]. This provided an estimate of the crystalinity change and amorphous region loss within the polymer matrix to determine the pre-degradation behaviour. Replicates of the specimens were tested for haze (%) and light transmission (%) and the average of the readings were reported.

3 RESULTS AND DISCUSSION

Table 1 depicts a sample of the mechanical properties estimated in this work. Typically, stress values are not related to the waste content in the blend [17]. Due to polymer films processing conditions, the change in the amorphous region affects the stress at break, which results from the heating cycles at compounding stages [18]. Stress at break was measured for the LLDPE and its blends and was compared with PSW. The effect of the compounding technique on stress at break, as well as, additive content was evident in the stress at break values (Fig. 1). SSE compounding gave the highest stress at break values for the LLDPE and PSW blends with and without PDQ-M. On the other hand, LLDPE/TPS blend exhibited the highest stress when compounded using TSE. Overall, the presence of the additives has affected the stress at break of the LLDPE polymer only. The PDQ-M additive allowed LLDPE blends to withstand the highest stress at break values compared to the pure LLDPE. The PSW, PSW/PDQ-M and PSW/TPS samples showed the same mechanical behaviour, i.e. they have similar stress at peak values for all compounding techniques. In general, the reported stress at break for pure LLDPE is several folds higher than that reported earlier in literature. TPS is a low-cost biodegradable polymer that can offer undesirable mechanical properties [19]. Adding TPS to the blends has been found to reduce the modulus and tensile strength of the Young but generally speaking this impact becomes greater as the TPS content increases [20]. The TPS sample reported the highest stress at break value (373 MPa) compared to all samples tested. This indicates that on a molecular level, the crystallinity of the TPS formulation is higher than the other examined blends, which is due to it exhibiting a higher stress at break as observed in Fig. 1. Again, the reported TPS stress at break values were several folds higher than those reported in literature [20], [21]. A study of particular interest is that of Shan et al. [22] which investigated the change in tensile stress over strain for a mixture of LDPE and two types of PP, namely homogeneous PP (PPH) and PP block-copolymer (PPC), over the full range of blend composition. The blends were compounded in a TSE and then injection moulded into test specimens. The mechanical properties of the blends were reported to show that pure LDPE had the lowest stress point (about 11 MPa) compared to its blends with PP.

Table 1: Sample of the mechanical properties estimated in this work showing the mechanical tensile pull properties of the blow film samples.

Material/Blend (wt.%/wt.%)	Young's Modulus (MPa)	Stress at break (MPa)	Strain at break (%)	Strain at break (N)
LLDPE (100)	5048 ± 1783.94	291 ± 11.49	922 ± 20.30	29.1 ± 1.15
LLDPE/PDQ-M (97/3)	3333 ± 740.14	305 ± 15.12	939 ± 28.44	30.5 ± 1.51
LLDPE/TPS (97/3)	3697 ± 254.46	276 ± 6.31	887 ± 21.51	27.6 ± 0.63
PSW (100)	3891 ± 199.90	194 ± 5.99	782 ± 20.80	19.4 ± 0.60
PSW/PDQ-M (97/3)	3718 ± 269.67	186 ± 15.25	749 ± 50.54	18.6 ± 1.52
PSW/TPS (97/3)	3951 ± 181.57	184 ± 7.02	752 ± 45.17	18.4 ± 0.70

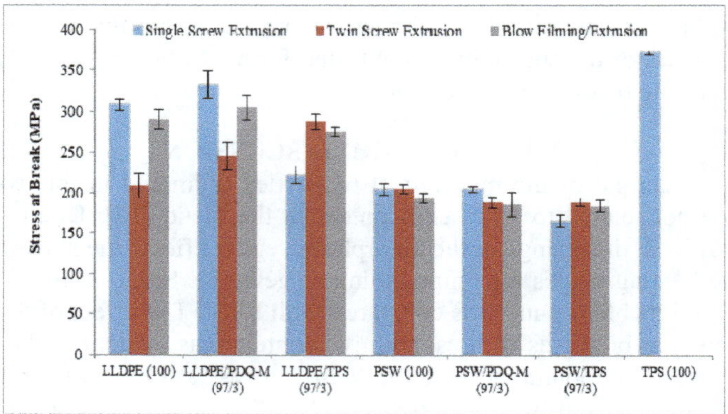

Figure 1: Stress at Bre||AK (MPa) measured for the compounded LLDPE and PSW films without and with TPS and PDQ-M.

The mechanical behaviour of the graphite nanophalates/LDPE manufactured with compression moulding and blown film extrusion technologies, was investigated by Carotenuto et al. [23]. The pure LDPE was found to be capable to stand extension up to 600% by activating necking and cold-drawing mechanisms. In BF, mechanical properties in the machined direction (MD) and the transverse direction (TD) were tested. In the transverse direction the unreinforced LDPE in the blown extrusion process exhibited a behaviour that is very similar to that observed for compression moulded samples. The polymeric chains' orientation effect was clearly visible in the Machined Direction (MD) sample curve, where LDPE behaved more like a brittle material, hitting higher tensile strength values (22 MPa) but a lower maximum pressure. However, the maximum strain reached only 300% which was almost twice for previously tested LDPE. Similar results were reported by Guichon et al. [24]. Another important consideration was the Young's modulus, which was not influenced by the transverse and the machined direction (180 MPa for TD and 185 MPa for MD), which means that the chain orientation mainly affected the necking and re-crystallisation phases and not the elasticity of polymers. Mechanical properties in terms of compressive response of LDPE were investigated across a range of strain rates and temperatures [25]. The LDPE plate samples were machined in the through-thickness direction, with a few experiments conducted in the two orthogonal directions. The LDPE stress–strain curve revealed an initial linear elastic region followed by a non-linear transition to yield, followed by strain softening and strain hardening afterwards. As predicted in most polymers, LDPE is showed an increase in stress with higher strain rate or lower temperature. LDPE stress-strain response was qualitatively similar to Omar et al. [26] LDPE response.

Strain at break (%) was estimated and depicted in Fig. 2. A clear variation is seen in relation to the additive content in the measured stain at break values for all the LLDPE formulations using the different compounding techniques. Furthermore, pure LLDPE and LLDPE-PDQ-M samples compounded in SEE exhibited the highest change in terms of strain when compared to their compounded samples using TSE and BF. This is in contrast to LLDPE-TPS sample compounded using SSE where a clear drop is noted in its strain value. This can be attributed, to the immiscibility on a physical level of the polymeric blend using this compounding technique. For the PSW and its blends it can be seen that similar strain values were maintained for all compounding techniques. This shows that an amount of 3 wt% of additive could be added to PSW samples with no effect on the product integrity. All of the samples tested in this study revealed higher strain at break values than TPS, indicating superiority of the blends over the TPS material. All of the PSW samples measured in this study showed relatively high strain at break values in comparison to the pure LLDPE and its blends, proving clear superiority of the waste blends over the pure polymer from an economical and environmental point of view.

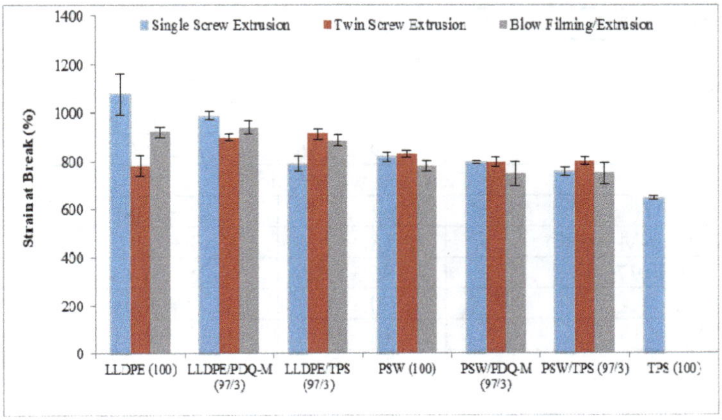

Figure 2: Strain at break (%) measured for the compounded LLDPE and PSW films without and with TPS and PDQ-M.

Colour parameters were measured and summarised in Table 2. The effect of compounding techniques on the blends in terms of the difference in the measured colour parameters was evident in SSE samples. LLDPE/PDQ-M blend compounded in SSE were lightest in colour and more yellow compared to LLDPE and LLDPE-TPS. Using the same compounding technique also, PSW had the highest lightness value compared to its blends. Amongst all compounded samples using SSE, TPS had the highest lightness, while L* values of its blends declined by increasing LLDPE and PSW content (97 wt.%) in the films. Lightness recorded for pure LLDPE, PSW and their blends compounded in TSE and BF were relatively within the same range (Table 2). Overall, the LLDPE and PSW films did not have the same L* and a* values in all compounding techniques. The same was also observed for their blends with PDQ-M and TPS. The a* values which stand for redness (or greenness) in colour space were insignificant, and even for films compounded in TSE and BF they were near zero. For b* values, PSW/TPS blend and PSW films had the highest yellowness levels in SSE and BF, respectively. The LLDPE/PDQ-M blend compounded in TSE had the most yellowness level amongst all compounded films. In general, lightness (L*) was in line with the general colour measurement theory, since transparent and achromatic films should have high lightness and minimum colour space values of a * and b*. Al-Salem et al. [27], [28] also tested colour parameters of control unexposed LLDPE/Waste (100/0) sample extruded using a single screw extruder (L/D=30) and film blowing machine. The reported L*, a* and b* values were 29.0, –0.4, and –3.5, respectively. These were slightly different from those of the pure LLDPE colour parameters reported in the study. Al-Salem et al. [28] also concluded that there was no change in colour parameters with respect to DHT.

Table 2: Colour parameters compounded LLDPE and PSW films (with and without TPS and PDQ-M additives).

Material/Blend (wt.%/wt.%)	L*	a*	b*
Single screw extrusion			
LLDPE (100/0)	35.06 ± 0.55	-1.16 ± 0.07	0.25 ± 0.28
LLDPE/PDQ-M (97/3)	44.50 ± 1.74	-1.43 ± 0.34	0.90 ± 0.43
LLDPE/TPS (97/3)	40.29 ± 0.33	-1.22 ± 0.13	0.28 ± 0.14
PSW (100/0)	37.92 ± 0.67	-1.31 ± 0.12	0.57 ± 0.41
PSW/PDQ-M (97/3)	23.93 ± 0.43	-1.47 ± 0.13	0.18 ± 0.05
PSW/TPS (97/3)	34.60 ± 0.95	-1.17 ± 0.41	3.48 ± 0.37
TPS (100/0)	72.49 ± 0.16	-2.33 ± 0.01	-3.91 ± 0.07
Twin screw extrusion			
LLDPE (100)	27.64 ± 0.39	0.22 ± 0.06	2.40 ± 0.22
LLDPE/PDQ-M (97/3)	28.58 ± 0.16	0.40 ± 0.05	3.59 ± 0.18
LLDPE/TPS (97/3)	23.86 ± 0.06	-1.25 ± 0.13	0.14 ± 0.22
PSW (100)	31.50 ± 0.11	-0.32 ± 0.04	0.56 ± 0.13
PSW/PDQ-M (97/3)	27.46 ± 0.12	-0.31 ± 0.11	0.37 ± 0.12
PSW/TPS (97/3)	32.37 ± 0.15	-0.68 ± 0.07	-0.72 ± 0.19
Blow filming/extrusion			
LLDPE (100)	30.90 ± 0.18	-0.09 ± 0.03	0.44 ± 0.14
LLDPE/PDQ-M (97/3)	29.73 ± 0.11	0.06 ± 0.03	1.36 ± 0.08
LLDPE/TPS (97/3)	26.32 ± 0.13	-0.42 ± 0.09	0.63 ± 0.32
PSW (100)	27.54 ± 0.18	0.22 ± 0.09	2.08 ± 0.64
PSW/PDQ-M (97/3)	31.69 ± 0.07	0.27 ± 0.01	1.57 ± 0.02
PSW/TPS (97/3)	31.65 ± 0.17	0.01 ± 0.00	0.66 ± 0.63

Light transmission (%) values were measured for LLDPE and PSW blends and are shown in Fig. 3. All pure samples and their blends showed similar transparency characteristics in comparison with TPS. Samples containing 97 wt.% LLDPE and PSW did not lose their transparency more rapidly because of the low percentage of additives in the blends. It is generally ideal for the packaging industry to manufacture films with low hazel and high gloss when the product does not need light protection [29]. Factors that can influence haze are the degree of crystallisation and/or crystal structure, and the time of polymers relaxation [30]. Higher relaxation time allows the crystallisation to occur under the influence of stress elongation, allowing the crystalline structures to be small, thin and oriented (lower haze and higher gloss) [31]. In addition, the loss of amorphous regions in polyolefins leads to hazy specimens; this was reported for TPS and its blends (LLDPE/PDQ-M and PSW/TPS) and is shown in Fig. 4. Pure LLDPE and its blends showed comparatively similar trend amongst all three compounding techniques indicating the dependency of haze (%) on the additive presence and the compounding technique applied. The LLDPE blends showed a clear shift in haze values when compared to the pure LLDPE. The same was also observed for PSW and its blends (PSW/PDQ-M and PSW/TPS). The virgin LLDPE proved tested by Al-Salem et al. [27] to be hazier its other blends with waste. The reported Haze% and Light Transmission% of unexposed virgin LLDPE (100) were 42% and 91.5%, respectively. The Haze% value was higher than that reported in the study for all LLDPE samples. The light transmission on the other hand was in agreement with that reported by Al-Salem et al. [27] (91.5%). Al-Salem et al. [27] also reported a notable decrease in haze measurement as the waste content increased and light transmission showed a gradual decrease with exposure time. In addition, Al-Salem et al. [28] reported a decrease in Haze% and light transmission% of the virgin LLDPE/waste blends (100/0, 90/10, 75/25, 50/50, 75/25, and 10/90), compounded using single screw extruder (*L*/*D* =30), with the increase in waste content due to the lack of miscibility between constituting polymers. The sensitivity of colour parameters to waste content was not evident in this study as seen in Fig. 4. The haze% and light transmission% were measured for unexposed LLDPE/waste (100/0) and (0/100) blends extruded, and blown using a single screw extruder (Tecnova, L/D=30) as part of the study conducted by Al-Salem et al. [8]. The study reported LLDPE/waste (100/0) haze% and light transmission% values of 41.27 ± 3.1 and 91.48 ± 0.09, respectively. The reported haze% and light transmission% values for LLDPE/waste (0/100) blend were 24 ± 0.32 and 90.60 ± 0.08, respectively.

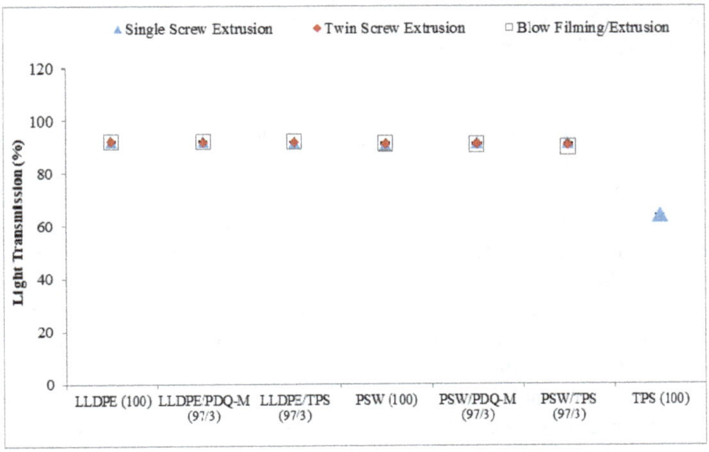

Figure 3: Light transmission (%) of compounded LLDPE and PSW films without and with TPS and PDQ-M.

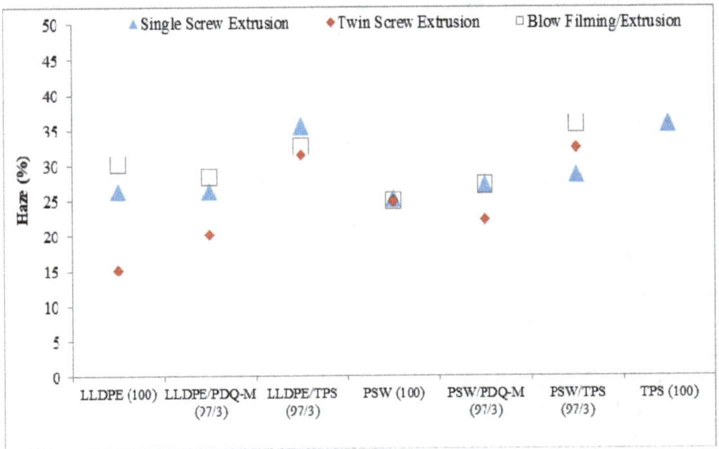

Figure 4: Haze (%) of compounded LLDPE and PSW films without and with TPS and PDQ-M.

It should be highlighted at this stage that oxo-biodegradables were criticised as of late for the possibility of environmental pollution (due to chemicals/toxics release to surrounding environment) and contributing to the overall microplastic (MP) release. These issues are related to oxo-biodegradables fragmentation in open environments and landfill conditions, as previously depicted in the EU Directive 94/62/EC [32]. Oxo-biodegradables are also known to be non-compostable materials, unlike hydro-biodegradables investigated in the research presented in this article. Combined with the presented efforts, and following previous recycling and biodegradation of plastics efforts in the region and its impact on the environment [9], [27], [33]–[36] the promotion of hydro-biodegradable plastics presents itself as a viable environmental solution. The execution of landfill burial experiments and following the same methodology for investigating the impact of landfill conditions on hydro and oxo-biodegradables presented elsewhere [8], [36] will also be part of this major research project in the near future that can deliver a more realistic assessment of environmental impact of such materials.

4 CONCLUSION

Formulated blends of virgin linear low density polyethylene (LLDPE) and plastic solid waste (PSW) with pro-degradant additive and starch based resin, were studied for their mechanical integrity and physical characteristics. All blends were in-line with past efforts characterising the mechanical integrity of market products which indicates their applicability to be mass produced as a biodegradable plastic for local markets. The compounding technique showed an immense impact on the stress at break of the blend. Single screw extrusion resulted in the highest value for the stress at break namely for the LLDPE and PSW blend. On the other hand, LLDPE/TPS blend exhibited the highest stress when compounded using TSE. It is therefore recommended to produce such blends with the aforementioned technique to meet certain demands and specifications of the market based on consumer demands. The change in mechanical properties could be attributed to the heating loop power and amount of shear applied due to the screw configuration. Starch based resin was also the haziest amongst sample produced in this work. However, the blending using twin screw extrusion showed that samples produced with this technique had the lowest haze (%) due to the high blending resulting from the twin screw configuration. There was no apparent impact with colour parameters light transmission between all samples.

ACKNOWLEDGEMENTS

The authors are grateful to the Kuwait Institute for Scientific Research (KISR) for supporting the activities of project EM065K. We acknowledge the financial support given by the Kuwait Foundation for the Advancement of Sciences (KFAS) towards the scientific mission to deliver the oral presentation of this communication.

REFERENCES

[1] Amass, W., Amass, A. & Tighe, B., A review of biodegradable polymers: uses, current developments in the synthesis and characterization of biodegradable polyesters, blends of biodegradable polymers and recent advances in biodegradation studies. *Polym Int*, **47**, pp. 89–144, 1998.

[2] Emadian, S.M., Onay, T.T. & Demirel, B., Biodegradation of bioplastics in natural environments. *Waste Management*, **59**, pp. 526–536, 2017.

[3] Chiellini, E., Cinelli, P., D'Antone, S. & Ivanova, S.V., Environmentally degradable polymeric materials (EDPM) in agricultural applications: An overview. *Polimery*, **47**, pp. 21–27, 2002.

[4] Orhan, Y. & Büyükgüngör, H., Enhancement of biodegradability of disposable polyethylene in controlled biological soil. *Int. Biodet. Biodegrad.*, **45**, pp. 49–55, 2000.

[5] Ohtaki, A. & Nakasaki, K., Ultimate degradability of various kinds of biodegradable plastics under controlled composting conditions. *Waste Manage. Res.*, **18**, 2000.

[6] Gomes, L.B., Klein, J.M., Brandalise, R.N., Zeni, M., Zoppas, B.C. & Grisa, A.M.C., Study of oxo-biodegradable polyethylene degradation in simulated soil. *Mat. Res.*, **17**, 2014. https://doi.org/10.1590/1516-1439.224713.

[7] Reddy, P.S., Subbaiah, G.V. & Subbaiah, H.V., Effect of plastic pollution on environment. *Environ. Sci.*, 2014. https://www.semanticscholar.org/paper/EFFECT-OF-PLASTIC-POLLUTION-ON-ENVIRONMENT-Reddy-Subbaiah/5c14101346a3684bcaa8289331a9c38abe5f8a6f.

[8] Al-Salem, S., Al-Hazza'a, A., Karam, H., Al-Wadi, M., Al-Dhafeeri, A. & Al-Rowaih, A., Insights into the evaluation of the abiotic and biotic degradation rate of commercial pro-oxidant filled polyethylene (PE) thin films. *J. Environ. Manage.*, **250**, 109475, 2019.

[9] Al-Salem, S.M., Sultan, H.H., Karam, H.J. & Al-Dhafeeri, A.T., Determination of biodegradation rate of commercial oxo-biodegradable polyethylene film products using ASTM D 5988. *J. of Polym. Res.*, **26**(7), p. 157, 2019.

[10] Al-Salem, S.M. et al., Thermal response and degressive reaction study of oxo-biodegradable plastic products exposed to various degradation media. *Int. J. Polym. Sci.*, 9612813, 2019.

[11] Montagna, L.S., de Camargo Forte, M.M. & Santana, R.M.C., Induced degradation of polypropylene with an organic pro-degradant additive. *J. Mater. Sci. Eng.*, **A3**, pp. 123–131, 2013.

[12] Ashori, A. & Nourbakhsh, A., Characteristics of wood–fiber plastic composites made of recycled materials. *Waste Manage.*, **29**, pp. 1291–1295, 2009.

[13] Vajna, B. et al., Complex analysis of car shredder light fraction, *Open Waste Manage. J.*, **3**, pp. 46–55, 2010.

[14] García M.T., Gracia I., de Lucas G.D.A. & Rodríguez J.F., Study of the solubility and stability of polystyrene wastes in a dissolution recycling process, *Waste Manage.*, **29**, pp. 1814–1818, 2009.

[15] Vilaplana, F. & Karlsson, S., Quality concepts for the improved use of recycled polymeric materials: A review. *Macro Mater Eng*, **293**, pp. 274–297, 2008.

[16] Al-Salem, S.M., Influence of natural and accelerated weathering on various formulations of linear low density polyethylene (LLDPE) films. *Mater. Design,* **50**, pp. 1729–1736, 2009.

[17] Elzubair, A. & Suarez, J.C.M., Mechanical behavior of recycled polyethylene/piassava fiber composites. *Mater. Sci. Eng. A*, **557**, pp. 29–35, 2012.

[18] Ha, K.H. & Kim, M.S., Application to refrigerator plastics by mechanical recycling from polypropylene in waste-appliances. *Mater. Des.*, **34**, pp. 252–257, 2012.

[19] Oliveira, T.A., Oliveira, R.R., Barbosa, R., Azevedo, J.B. & Alves, T.S., Effect of reprocessing cycles on the degradation of PP/PBAT-thermoplastic starch blends. *Carbohyd. Polym.*, **168**, pp. 52–60, 2017.

[20] Berruezo, M., Ludueña, L.N., Rodriguez, E. & Alvarez, V.A., Preparation and characterization of polystyrene/starch blends for packaging applications. *J. Plastic Film Sheet*, **30**, 2014.

[21] Pang, M.M., Pun, M.Y. & Ishak, Z.A.M., Thermal, mechanical, and morphological characterization of biobased thermoplastic starch from agricultural waste/polypropylene blends. *Polym. Eng. Sci.*, **54**, pp. 1357–1365, 2014.

[22] Shan, H., White, J.L. & DeGroot, A.W., Comparison of crystallization and melting characteristics of quiescent and melt-spun poly(ethylene-co-octene) copolymers. *Int. J. Polym. Anal. Charact.*, **12**, pp. 231–249, 2007.

[23] Carotenuto, G. et al., Mechanical properties of low-density polyethylene filled by graphite nanoplatelets. *Nanotech*, **23**, 2012. https://iopscience.iop.org/article/10.1088/0957-4484/23/48/485705/meta

[24] Guichon, O., Séguéla, R., David, L. & Vigier, G., Influence of the molecular architecture of low-density polyethylene on the texture and mechanical properties of blown films. *J. Polym. Sci. B Polym. Physc.*, **41**, pp. 327–340, 2003.

[25] Jordan, J.L. et al., Mechanical properties of low density polyethylene. *J. Dynamic Behavior Mater.*, **2**, pp. 411–420, 2016.

[26] Omar, M.F., Akil, H.M. & Ahmad, Z.A., Effect of molecular structures on dynamic compression properties of polyethylene. *Mater. Sci. Eng.*, **538**, pp. 125–134, 2012.

[27] Al-Salem, S.M., Abraham, G., Al-Qabandi, O.A. & Dashti, A.M., Investigating the effect of accelerated weathering on the mechanical and physical properties of high content plastic solid waste (PSW) blends with virgin linear low density polyethylene (LLDPE). *Polym. Test*, **46**, pp. 116–121, 2015.

[28] Al-Salem S.M., Al-Dousari, N.M., Joseph Abraham, G., Aromin D'Souza, M., Al-Qabandi, O.A. & Al-Zakri, W., Effect of die head temperature at compounding stage on the degradation of linear low density polyethylene/plastic film waste blends after accelerated weathering. *Int. J. Polymer Sci.*, 5147209, 2016.

[29] Alvarez, V.B. & Pascall, M.A., Packaging. *Encyclopedia of Dairy Sciences*, ed. J.W. Fuquay, Academic Press: New York, pp. 16–23, 2011.

[30] Zhang, X., Liu, X., Zheng, W. & Zhu, J., Regenerated cellulose/graphene nanocomposite films prepared in DMAC/LiCl solution. *Carbohydrate Polym.*, **88**, pp. 26–30, 2012.

[31] Borse, N. & Aubee, N., Predicting physical and optical properties of co-extruded blown films using design of experiment based model. Presented at *SPE ANTEC Orlando*, 2015.

[32] European Commission (EC), report from the commission to the European parliament and the council on the impact of the use of oxo-degradable plastic, including oxo-degradable plastic carrier bags, on the environment, 2018. http://ec.europa.eu/environment/circular-economy/pdf/oxo-plastics.pdf.

[33] Al-Salem, S.M., Uddin, S. & Al-Yamani, F., An assessment of microplastics threat to the marine environment: A short review in context of the Arabian/Persian Gulf. *Marine Environ. Res.*, **159**, 104961, 2020.

[34] Al-Salem, S.M., Uddin, S. & Lyons, B., Evidence of microplastics (MP) in gut content of major consumed marine fish species in the State of Kuwait (of the Arabian/Persian Gulf). *Marine Poll. Bullt.*, **154**, 111052, 2020.

[35] Al-Salem, S.M., Influential parameters on natural weathering under harsh climatic conditions of mechanically recycled plastic film specimens. *J. Environ. Manage.*, **230**, pp. 355–365, 2019.

[36] Al-Salem, S.M. et al., Thermal response and degressive reaction study of oxo-biodegradable plastic products exposed to various degradation media. *Int. J. Polym. Sci.*, 9612813, 2019.

COMPOUNDING AND PROCESSING HYDRO-BIODEGRADABLE PLASTIC FILMS FOR PLASTIC WASTE REDUCTION. PART II: THERMAL AND CHEMICAL PRINTING OF VIRGIN/WASTE POLYMERIC BLENDS

MASUMAH AL-QASSIMI, HAJAR JAWAD KARAM,
MAJED HAMEED AL-WADI & SULTAN MAJED AL-SALEM
Environment & Life Sciences Research Centre (ELSRC), Kuwait Institute for Scientific Research (KISR), Kuwait

ABSTRACT

In an effort to develop a biodegradable blend that could withstand arid environmental conditions; and standardise the use of biodegradable plastic films originating from hydro-biodegradable blends, this project was initiated by the Kuwait Institute for scientific Research. In this work, linear low-density polyethylene (LLDPE) was compounded with a 3 wt.% thermoplastic starch (TPS) resin and by the same amount with a pro-degradant chemical, using three different techniques that are standardised in the industry of plastics conversion. These are extrusion by single screw compounding, twin screw compounding and blown-filming machine as well. The materials were studied in comparison to the virgin LLDPE, as well as, formulated blends with plastic solid waste (PSW) materials, also in an effort to determine the impact such chemicals and pro-degradants have on the municipal solid waste (MSW) stream. Firstly, mechanical properties were studied using a universal testing machine by applying ASTM 882-12 and compared to other market grades available within localities. Furthermore, the thermogravimetric behaviour and infrared spectroscopy (IR) were established and have resulted in a clear shift within the onset and end set temperatures, in comparison to other commercial grades. This study establishes a baseline for compounds that could be used in the future as a biodegradable plastic film in an effort to combat plastic solid waste (PSW) accumulation in developing countries.

Keywords: biodegradation, polyester, starch, polymer waste, waste management.

1 INTRODUCTION

Packaging is a process that encloses or covers goods intended for shipment, storage, sale and use. Polyethylene (PE) is the commonly used polymer in packaging. It is converted into packaging materials by thermal processes such as film blowing, sheet extrusion, or extrusion coating [1]. The packaging sector is the largest consuming sector by demand for all polymer types and resin. Fig. 1 shows the global demand for plastics by sector where packaging, building and textiles represents the major share of the market by 34%, 17% and 10%, respectively [2]. The increase in dependency on plastics has led to an increase in the amount of plastic solid waste (PSW) generation, which is estimated to be 260 million tonnes per annum on a global basis [3].

The properties of products play an important role in improving the quality and shelf-life of the food in terms of food packaging and other products where plastic covers are typically used [4]. A key requirement for food packaging materials is that they should not emit substances that interfere with the product, i.e. cause odour and taste. EQUATE Petrochemical Company in the State of Kuwait currently produces approximately 825 ktpa of linear low-density polyethylene (LLDPE) and high-density polyethylene (HDPE) with an expected increase in market demand in the coming years. Plastic converters that make shopping bags, trash bags, agricultural mulches, greenhouse covers and many others use a significant portion of these polymers locally most of which end up as waste. The generation of PSW in Kuwait is estimated at around 150 to 200 ktpa and is predicted to grow as a result of population growth and high living standards in Kuwait. The predominant technique of disposal of solid waste (SW) in Kuwait is landfill burial and with PSW taking very long time to effectively degrade in the environment. PSW accumulation and management is perceived to be one of the main issues facing both government and private sector parties. Recent estimates have shown that Kuwait produces 825 ktpa of PE and imports more than 25 ktpa of virgin resin most of which are used in the packaging industry and the production of bags [5].

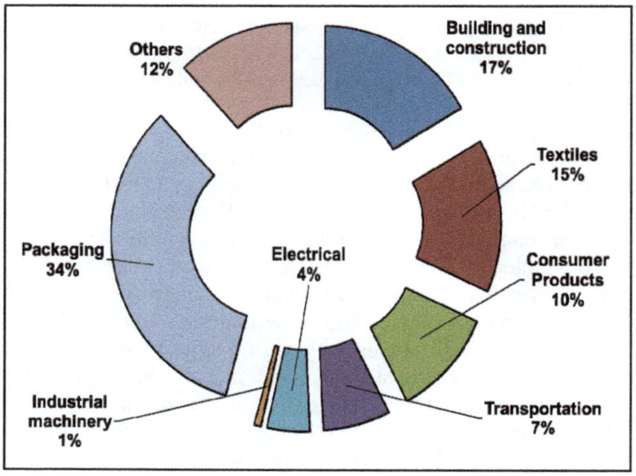

Figure 1: Plastic resin global demand by sector. *(Source: Redrawn from Parker, 2018 [2].)*

In this work, linear low-density polyethylene (LLDPE) was compounded with a 3 wt.% thermoplastic starch (TPS) resin and by the same amount with a pro-degradant chemical, using three different techniques that are standardised in the industry of plastics conversion. These are extrusion by single screw compounding, twin screw compounding and blown-filming machine as well. The materials were studied in comparison to the virgin LLDPE, as well as, formulated blends with plastic solid waste (PSW) materials, also in an effort to determine the impact such chemicals and pro-degradants have on the municipal solid waste (MSW) stream.

2 EXPERIMENTAL

LLDPE (grade EFDC-7050) was used as received in a translucent pellet form with a reported density of 0.918 g cm^{-3} [1]. The LLDPE polymer was compounded with a 3 wt.% thermoplastic starch ECOVIO (TPS) resin and by the same amount with a PDQ-M Willow Ridge Plastics (Inc.) pro-degradant chemical, using three different techniques that are standardised in the industry of plastics conversion. These techniques are namely, (a) single screw extrusion, (b) twin screw extrusion, and (c) blow-filming which are the most common among commercial film production lines, and have been chosen to study the effect of processing and melting on the materials. Throughout the rest of this paper, these techniques will be coded as follows: a) SSE, (b) TSE, and (c) BF, respectively. The PDQ-M pro-degradant chemical was used as received in the form of tan pellets and a melting point (T_m) equal to 110°C. The thermoplastic starch (TPS) was used as white translucent pellets and was certified by the manufacturer in accordance with EN17033 with a T_m equal to 120°C. SSE was performed using a Dolci extruder (L/D ratio 1/28) using 10 kg as master batch mix. The blends with TPS were conducted in the same mix ration as the oxo-additives, to have a comparable baseline between the materials. Mixing with palletized PSW was also conducted in the same blend ration of 3 wt.%. Furthermore, BF was conducted with the same formulations aforementioned using a Windsor (monolayer) blown-film coupled with an L/D (1/28) single screw extruder operated under 6 bars of pressure with an RPM of maintained between 75 and 80. TSE was performed using a Kung Hsing (1:28, L/D ratio, 6 bars) operated in co-rotation with a maximum speed of 75 RPM maintained throughout the experimental work. The TPS samples were extruded in a standalone Brabender SSE which was used in the compounding process of the samples. All samples were extruded as sheets of 100 μm thickness meeting standard market specifications and as an average of major film applications. The samples appeared consistent with no visual fractures or crazing and were comparable to each other by touch. The mechanical properties were studied following test protocol of ASTM 882 using a Tinius Olsen UTM machine as depicted previously in Al-Salem et al. [1]. Variation in processing and cutting was negligible over the length of some samples between the width and thickness, accounting for less than 1% of the total number of samples measured. The variation for the samples was not greater than ± 0.05 mm for width and ± 0.02 mm for thickness, and a constant width of 12.50 mm and 0.10 mm for thickness was taken. Thermogravimetric analysis (TGA) was conducted for all compounded compounded LLDPE and PSW films without and with TPS and PDQ-M under nitrogen atmosphere in a Mettler-Toledo TGA Unit coupled with a StarE software using 7.5 mg sample mass and at a heating rate of 15°C min^{-1} from room temperature to 600°C. Following previously conducted work [6], [7], the samples were thermally monitored via infrared spectroscopy using Fourier transform infrared spectroscopy (FTIR) in attenuated total reflectance (ATR) mode. The film samples were mounted on the crystal of the machine's holder, and the spectra after each sample unloading was taken. Background spectra were collected before each measurement. An initial frequency range of 500 to 5000 cm^{-1} with an average 32 scan of a resolution of 4 cm^{-1} was used at room temperature. The ketone carbonyl bond at 1714 cm^{-1} was considered and its bond index (KCBI) was estimated as the ratio of the carbonyl group, band and that of the PE band thus

$$KCBI = \frac{I_{1714}}{I_{1470}}. \tag{1}$$

The 1470 cm^{-1} absorbance is taken as the PE band. Upon studying the evolution of the carbonyl band, the other component indices were calculated and the main products according to literature [8] are namely: Lactone (L), Ester (E)/Aldehydes (A), Carboxylic Acid (CA), and Vinylidene (V). An index was allocated to each of these products, which was determined to measure the degree of thermal degradation in relation to the PE band reference point, bearing in mind that reference bands could vary with the type of polymer film and the type of test conducted. The target bands may differ, hence at first the peaks must be tracked and calculated using the FTIR results of the control sample

$$LBI, EBI/ABI, CABI, VBI = \frac{I_n}{I_{1470}}, \tag{2}$$

where n stands for I1714, I1770, I1733, I1698, and I888 which are which are associated with band absorbance at 1714, 1770, 1733, 1698, 888, respectively following previously depicted methodology shown elsewhere [9].

3 RESULTS AND DISCUSSION

The extrusion profile with respect to temperatures of various temperature zones against for all three compounding techniques showed a typical processing conditions trend that was reported previously by other authors [1], [9]. The relationship between the polymer structure and its mechanical behaviour should not be disregarded, and several studies have been carried out over the last few years to investigate the impact of molecular structures on the overall characteristics of PE, in particular its mechanical properties [10]–[12]. LLDPE is typically produced by co-polymerisation of ethylene and such higher alpha olefins, including butane, hexane or octane, at lower temperatures and pressures. The copolymerisation process gives LLDPE a narrower molecular weight distribution, linear structure and significantly different rheological properties than traditional LDPE polymer. The sample preparation technique was taken into account when comparing the LLDPE data from this study with data from other PE studies. One of the earliest studies conducted on LLDPE was that of Dumoulin et al. [13] who examined the tensile properties of two series of LLDPE with isotactic PP. The blends were prepared using TSE at 245°C. The composition dependence of Young's modulus and of the strain at break versus concentration of PP in LLDPE blends was reported. The strain at break of LLDPE-1/PP (100/0) and LLDPE-2/PP (100/0) blends were 6.5% and 8.5%, respectively. The Young's modulus measured for LLDPE-1/PP (100/0) and LLDPE-2/PP (100/0) were approximately 0.30 and 0.13 GPa, respectively. Change in mechanical properties of LLDPE polymer and molecular structure due to degradation were followed in depth by Naddeo et al. [14]. Mechanical parameters were also correlated to exposure time and carbonyl index. The LLDPE pellets were moulded in a hot press at 240°C to form films of 130 μm thickness. The stress–strain curve of the initial LLDPE sample before being exposed to accelerating weathering derived the mechanical properties of interest namely: Elastic modulus (120 MPa), stress at yield point (8.3 MPa), stress at break (9 MPa), strain at break (1690%) and toughness (323 MPa). The mechanical properties in terms of Young's modulus, yield stress, impact strength, and elongation at break of pure and zeolite filled LLDPE blends were studied by Biswas et al. [15] using a UTM of Instron 4465, at 25°C and 30% humidity. The pure LLDPE and its blends were prepared using a Brabender twin-screw extruder (PL 2000) with L/D screw dimension of 16 at temperature gradient of 170°C in the feeding zone, 180°C in the compression zone, 190°C in the metering zone and 200°C at die. The screw's rotation speed was kept between 60 and 70 rpm. Two specimen types, dumbbell and film, were prepared for the mechanical properties' measurement at different crosshead speeds. For the film specimens (dimensions: 15 x 0.4 x 165 mm^3) in particular, the initial grip distance was between 50 and 500 mm /min. The stress–strain curves of the pure LLDPE film showed that elongation at break for 5 mm/min exceeded 1000% (i.e. sample can be elongated up to 500 mm) and a similar behaviour was observed at 500 mm/min. In addition, for the pure LLDPE, the Young's modulus obtained for 50 and 500 mm/min strain rates was 120 and 127 MPa, respectively. The yield stress measured for the pure LLDPE at strain rates of 50 and 500 mm/min were 6.8 and 9.5 MPa, respectively. The elongation at break for the pure LLDPE was 200% and 1000% (i.e. machine can elongate the sample up to 500 mm) at strain rates of 50 and 500 mm/min, respectively.

The comprehensive literature review conducted on the mechanical properties of LLDPE, LDPE, and HDPE and their techniques of compounding provided a solid basis for comparison of the films examined in this study. The films compounded were all comparable to market standards which renders them applicable for use as a carrier film and packaging product [1]. The polymer blends compounded using the SSE, TSE and BF showed slight variations in the Young's modulus (Fig. 2). The compounding method and the incorporation of TPS and PDQ-M additives into the blends altered the mechanical properties of the samples in comparison to the pure LLDPE and PSW. The highest calculated Young's modulus value for the pure LLDPE was 5048 MPa compounded using Blow-filming/extrusion which was much higher than the other LLDPE blends with and without additives (Fig. 2). The highest calculated Young's modulus value for PSW (100) compounded using twin screw extrusion was 4547 MPa which was slightly higher than that reported for its blends with TPS and PDQ. This showed that their relative compounding techniques allowed the pure blends (100 wt.%) to exhibit higher stiffness and less rapid deformation. In general, the Young's modulus reported in the study are several folds higher than those reported in literature. This could be attributed to the high content of plasticisers in LLDPE and PSW which gave increased flexibility to the compounded films.

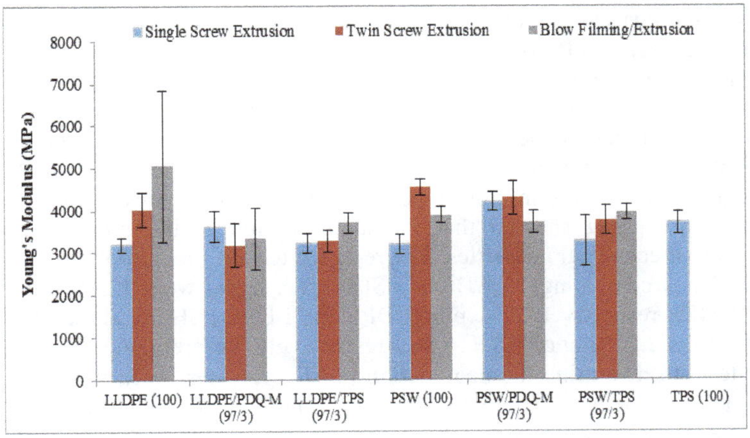

Figure 2: Young's modulus (MPa) measured for the compounded LLDPE and PSW films without and with TPS and PDQ-M.

The estimated Young's modulus value of the TPS compounded in single screw extrusion was 3696 MPa. The thermal and mechanical properties of a comparative TPS commercial grade blend with PP were investigated by Pang et al. [16]. Sheet extrusion was performed in a Berlyn model single-screw extruder with an L/D ratio of 31:1 and processing temperature range of 150–170°C, and screw speed of 50 rpm. Tensile properties of dumbbell specimens were determined with an Instron model 3366 tester according to ASTM D 638, at a cross-head speed of 50 mm/min. The tensile properties of the extruded pure TPS (100) sheets showed the lowest tensile Strength (5–7 MPa), Elongation at break (6%–16%) and relatively high Young's modulus (540–660 MPa) values compared to the other blends. This is due to the high starch content in the formulation (biobased content above 50%) compared to NTS/PP, AWS/PP, and AWT/PP blends (biobased content 35%–37%). This finding is consistent with previous research, where the introduction of starch granules to synthetic polymers influenced the mechanical properties by reducing tensile strength and elongation at break, while the Young's modulus increased as the starch content increased due to the stiffing effect of the granules [17]. In addition, according to Azhari and Wong [18] a high content of starch in the sample can cause poor dispersion as starch granules agglomerate, leading to a reduction in mechanical properties. Polystyrene/TPS blends were prepared by melt mixing and characterised by TGA, FTIR and tensile tests (strength, modulus, elongation at break and tenacity) [19]. PS/TPS blends were prepared using an intensive Brabender type mixer with two counter-rotating roller rotors at a mixing temperature of 155°C, speed rotation of 150 RPM and mixing time of 10 min. Films were prepared by compression moulding with a thickness of 300–500 mm. For TPS (0/100) blend, the stress–strain curve, conditioned at 30% Relative Humidity (RH), showed a linear elastic behaviour and a subsequent plastic deformation but without strain hardening. The curves obtained from the tensile tests were used to measure the mechanical properties of TPS at 30% RH and 25°C. Tensile Strength of 9.5 MPa, Elongation at break of 3.4% and Young's modulus of 790 MPa were measured at 30% HR. The effect of relative humidity on the mechanical behaviour was studied further at 60% and 75% RH to test the environment impact on the mechanical behaviour of TPS. Humidity has been found to have a major impact on TPS' mechanical properties as it serves as a plasticiser for TPS, lowering the modulus and tensile strength of its blends and increasing the elongation at breakage. The impact of TPS on lowering Young's modulus was mainly evident in all the LLDPE blends as see in Fig. 2.

The degradation of pure LDPE and its blend with a starch-based pro-degradant additive (PSH) (25 wt.%) under various degrading media was previously studied [20]. The LDPE/PSH blend was compounded in a Brabender Plasticorder (Mixer 50E) for 10 min at temperature of 1408°C and mixing speed of 60 rpm. The stress–strain plots for the LDPE/PSH before aging showed that the incorporation of the pro-degradant additive caused a drop in mechanical properties. The reported tensile strength of the pure LDPE and LDPE/PSH blends containing 25 wt% of PSH before aging were 15.30 ± 0.09 MPa and 11.40 ± 0.10 MPa, respectively. The pure LDPE and LDPE/PSH blend recorded elongation at break of 338 ± 2.12% and 118 ± 6.8%, respectively, before aging. The application of weathering tests to determine the degradation profile polymers used in plastic production has been studied to a great extent in literature [21]–[24]. The mechanical and physical properties of PSW in the form of discarded plastic films were tested along the pure LLDPE polymer [24]. Dry blends of LLDPE and PSW were extruded, and blown using SSE (L/D=30). Pure LLDPE/PSW ratio of the compounded blends were 100/0, 95/5, 90/10, 75/25, 50/50, 25/75, 10/90 and 0/100. The mechanical behaviour of pure LLDPE (100) and PSW (100) blends in terms of the Young's modulus, stress and elongation at break was assessed and compared to the mechanical assessment conducted in this study. The Young's modulus measured for the pure LLDPE and the PSW blends was 21 and 72 MPa, respectively. The PSW blends also showed a decreasing trend in the modulus as the waste content increased signifying that the plastic film waste acts as a degrading agent to the pure LLDPE. The pure LLDPE and PSW samples had a relatively similar stress at break values of approximately 19 MPa and 18 MPa, respectively. The pure LLDPE an PSW samples had a strain at break values of 1478% and 500%, respectively. From the study a clear decrease in mechanical properties was observed as a result of both the waste material and natural weather exposure. Al-Salem et al. [23], [24] also evaluated mechanical and physical (optical) properties of pure LLDPE and its blends with PSW under accelerated weathering. They further investigated in depth the effect of die head temperatures (DHTs) at compounding stage on the degradation of the compounded blends [24]. Samples properties were tested at three DHTs in the extruder section: 175°C, 185°C and 195°C and those reported at 195°C in particular were used in comparison to the current study for samples compounded using BF/ESS DHT of 190–200°C. The Young's modulus and stress at break values reported for pure LLDPE were similar at all DHTs reaching values of a round 62 MPa and 18 MPa, respectively, at DHT=195°C. The strain at break reported at 195°C DHTs was 1500%. On the other hand, the Die Head Blown-film method used by the extruder altered mechanical properties of the samples when waste fractions were added to the blends in comparison to the pure LLDPE. The Young's modulus and stress at break values reported for LLDPE/PSW (10/90) at DHT=195°C were 20 MPa and 14 MPa, respectively. A strain at break value of 400% was reported at 195°C DHTs for the LLDPE/PSW (10/90) blend.

For the SSE, TSE and BF techniques (used in this work) the ranges of Young's Modulus measured for the pure LLDPE, LLDPE/PDQ-M (97/3), and LLDPE/TPS (97/3) were 3180–3620, 3176–4027, and 3333–5048 MPa, respectively. For the PSW, PSW/PDQ-M (97/3), and PSW/TPS (97/3) blends compounded using SSE, TSE, and BF techniques the ranges of Young's Modulus measured were 3185–4210, 3766–4547, and 3718–3951 MPa, respectively. The tested films and PSW in particular were seen to have relatively high modulus values when compared to LLDPE films in literature and have surpassed the stiffness measurements required for the local market. The tested PSW films have shown potential for utilising the plastic waste in the local plastic production industry. The modulus is found to dependant on the individual modulus values of the materials used in the blend and is showing an increasing trend when certain polyolefin or polyester polymers are present in the blend in proportion to their amount [25]. In the study, Young's modulus was found not to be related to waste content in the blend especially when chemical additives were added due to the low TPS and PDQ-M percentage (3 wt.%).

All samples experienced a major weight loss stage in the TGA associated with the degradation of the synthetic polymer such as LLDPE. In the case of LLDPE and PSW blends with TPS two major weight loss stages were expected the first being that of the synthetic polymer and the second is typically due to the degradation of the starch. Also in this study, ATR-FTIR was used as a qualitative assessment method to aid in polymer identification based on their specific absorption bands. Each FTIR spectrum of LLDPE and PSW virgin and compounded films were compared with reference spectra from the spectral database of the FTIR FDM Library. The program shows a variety of compounds correlated with correlation coefficients, of which the FTIR spectra matches the spectra of the analysed film sample. In order to determine the samples degree of degradation, the region of carbonyl group's absorption band (1680–1820 cm^{-1}) was investigated for all pre-degradable tested films. Martínez-Romo et al. [26] tested the changes in structural properties of HDPE, LDPE, bio-degradable polyethylene (PE-BIO), and oxo-degradable PE films without exposure to UVB radiation. The infrared spectra of the tested films showed the PE characteristics bands: 2772–3038 cm^{-1} (the stretching vibration of Carbon-Hydrogen (CH)), 1440-1490 cm^{-1} (shaking and rocking vibration of methylene (CH_2), 700–750 cm^{-1}, respectively [27]. The PE-BIO in particular showed peaks at 700–750 cm^{-1} and 1440–1490 cm^{-1} due to the polymers' linear structure, while for LDPE and PE-OXO, these same bands were divided because they represent branched polymers. The IR spectra of the PE-BIO films showed absorption band at 1740 cm^{-1} (stretching vibration of the carbonyl group (C=O) and PE-OXO film showed absorption band at 909 cm^{-1} (stretching vibration of the vinyl group (CH_2=CH)). Such functional groups are found in chromophores and are considered precursors of PE photochemical reactions [28]. The carbonyl and vinyl groups are considered the PE's key photo-oxidation products [29]; thus, the carbonyl index (ICO) and the vinyl index (IV) are parameters used to determine the effect of UV-B radiation in PE films [30]. The PE-BIO film IR spectrum without exposure to UV-B radiation presented the characteristic carbonyl group IR band (1640 cm^{-1}) which indicated that the PE-BIO content is oxidised during the manufacturing process.

FTIR spectra (see Fig. 3 for an example) showed that all LLDPE and PSW films compounded without TPS and PDQ-M using the three compounded techniques presented major characteristic PE absorbance bands that are located at 2916 cm^{-1} and 2846 cm^{-1} (C–H stretching asymmetric and symmetric vibrations), 1462 cm-1 (crystalline phase of CH_2 bend), and 719 cm^{-1} (CH_2 rocking vibrations). A weak absorption band at 1471 cm^{-1} (CH_2 scissoring vibrations) was also observed for all pre-degradable films suggesting that degradation is not occurring. The 1470 cm^{-1} and 798 cm^{-1} peaks are used to identify and quantify the presence of PE [31]. In general, the detected bands follow a polyethylene (PE) spectrum as previously depicted by past works [32]. In addition, the detected bands are in agreement with the FTIR spectra of unexposed [6], commercial LDPE grade material [33] and IR spectra of the control LLDPE film blends [34] to predict the lifetime of oxo-degradable polyolefin and biodegradable polymer films. FTIR spectra obtained for LLDPE are comparable to published spectra for respective polymer blend [33]–[35].

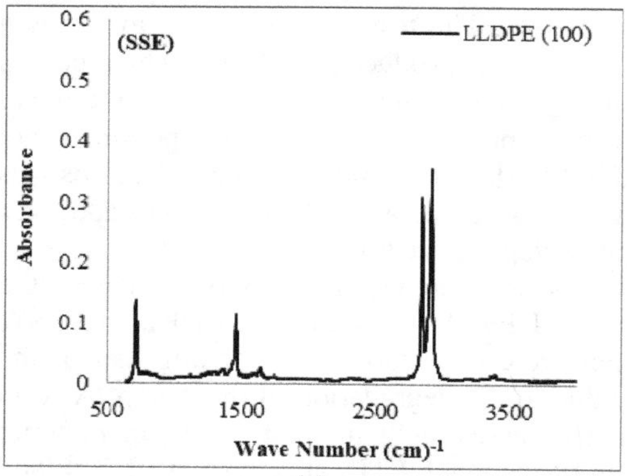

(a)

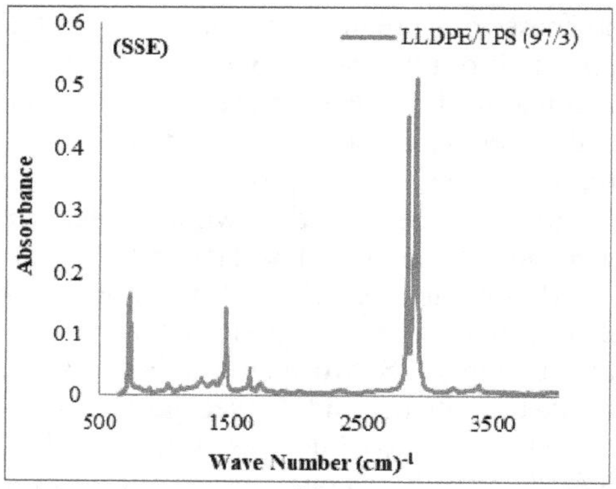

(b)

Figure 3: SSE samples showing the spectra obtained with ATR analysis for (a) LLDPE; and (b) LLDPE/TPS).

In general, compared to LLDPE and PSW the peak locations and peak shapes of the blends were alike, however, absorbance of LLDPE/TPS was higher than LLDPE and LLDPE/PDQ-M for single screw extrusion. In addition, it was also observed that absorbance of LLDPE was lower than that of its blends compounded with PDQ-M and TPS using blow-filming/extrusion. On the other hand, peak locations, peak shapes and absorbance of all films compounded in twin screw extrusion were alike for all tested films. The LLDPE films with PDQ-M also showed absorption band referent to PE 2916 cm^{-1} (CH_2 asymmetric stretch bonds), 1462 cm^{-1} (crystalline phase of CH_2 bend), and 719 cm^{-1} (crystalline phase of CH_2 rock). In general, very weak absorption band closer to 1715 cm^{-1} (stretching vibration of C-O groups) that indicates the presence of carbonyl (C=O) reflecting the esters of biopolymers and the possible oxidation of the LLDPE and PSW compounded films. In literature, a typical FTIR spectra of PE+AD (AD: oxo-degradable additive) blend will have main peaks corresponding to hydroperoxides (3200 cm^{-1}), alcohols (3416 cm^{-1}), Ketones (1715 cm^{-1}), aldehyde (1725 cm^{-1}), carboxylic acids (1710 and 1715), noncyclic/cyclic esters (1735 and 1785 cm^{-1}) and double bonds (909 cm^{-1}) [36].

The spectra of LLDPE/PDQ-M (97/3) and PSW/PDQ-M (97/3) had appreciable differences due to the low PDQ-M% in these blends. For TPS, weak absorption band at 2918 cm^{-1} (CH_2 asymmetric stretching vibration), strong absorption bands at 1711 cm^{-1} (C-O stretching vibration), 1267 cm^{-1} (C-O bonds symmetric stretching vibration of C-O bonds for alcohols), 1101 (bending vibration at the surface of adjacent hydrogen atoms on the phenyl ring bonds) and 727 cm^{-1} (bending vibration absorption of CH-plane of the benzene ring bonds) [37]. In addition, weak band at 1450 cm^{-1} (vibrations of the angular deformation of CH_2) was detected [34]. The absorption band at approximately 1016 cm^{-1} that is attributed to the extent of deformation and bending of C-O-C and OH groups of the starch structure (999 cm^{-1}) was detected. In addition, a large band in the 3350 cm^{-1} region that is typically associated with TPS structure was not found [38]. The PE-BIO film IR spectrum without exposure to UV-B radiation presented the characteristic carbonyl group IR band (1640 cm^{-1}) which indicated that the PE-BIO content is oxidized during the manufacturing process [26].

In particular, the ketone carbonyl and vinyliene indices are used to measure the degree of films degradation. The calculated KC bond indices from the FTIR spectra are found to be below 0.25. TPS on the other hand exhibited a much higher KC bond index of 3.2. In literature, films containing mainly PE were found to exhibit KC bond index around 1 which means that PE containing blends are more degradable compared to polypropylene (PP) [39]. An increase in the KC bond indices was observed for LLDPE and PSW blends with TPS and PDQ-M (97 wt.%/ 3wt.%) compared to LLDPE (100) and PSW (100). This is caused by the PDQ-M and TPS in the LLDPE and PSW films which cleaved the polymer chain, enhancing chain movement and leading to an increase in crystallinity [40]. The smaller double bond indices seen for some blends in comparison to others may suggest that the double bond was used in polymer crosslinking [40]. The effect of additives in increasing KC bond index was also observed in the study of Portillo et al. [36] where Carbonyl Index for PE+AD (AD is oxo-degradable additive) was found to be greater than PE. The KC bond index of unexposed LDPE was also reported in literature to be greater than HDPE and PE-OXO as it is a branched material with a greater number of tertiary carbon which makes it more prone to undergo photooxidation reactions [26]. The ketone carbonyl and vinylidene indices of LLDPE compounded in SSE were 0.11 and 0.12, respectively. The ketone carbonyl and vinylidene indices of LLDPE compounded in TSE were 0.07 and 0.08, respectively, and those of LLDPE compounded in BF/SSE were 0.06 and 0.075, respectively. The indices of LLDPE compounded in TSE and BF/SSE were in agreement to the ketone carbonyl and vinylidene indices of PE (0.08 and 0.09, respectively) [40]. The promotion of this research work, namely considering hydro-biodegradable plastics is considered to be quite topical as of late after the restriction of the use of oxo-biodegradable [41]. Following the research conducted by our group [42], [43] and the continuation of the research efforts in this communication, legislations are expected to presented and formulated by local authorities to dictate the use of both oxo and hydro-biodegradable plastics. These will also eliminate possible threats of fragmentation and environmental pollution in eth near future [42]–[45].

CONCLUSION

Various plastic blends encompassing pro-degradants and thermoplastic starch resin, where compared with virgin polyethylene, plastic solid waste and thermoplastic starch films. Compounding was achieved using single, twin and blow film extrusion to produce 100 μm thinness films which were tested for mechanical integrity and thermal properties. Thermogravimetry showed a single stage loss to all materials with the exception of the starch based compound indicting loss of polyester and starch. On the other hand, the samples showed minimal changes with respect to compounding technique and chemical fingerprinting using infrared spectroscopy. Furthermore, the materials were of acceptable market properties that allowed comparison of mechanical properties to typical plastic film products. This points towards the fact that biodegradable plastic waste blends could be utilised in the future as a standalone product to reduce dependency on virgin plastic materials.

ACKNOWLEDGEMENTS

The authors are grateful to the Kuwait Institute for Scientific Research (KISR) for supporting the activities of project EM065K. We acknowledge the financial support given by the Kuwait Foundation for the Advancement of Sciences (KFAS) towards the scientific mission to deliver the oral presentation of this communication.

REFERENCES

[1] Al-Salem, S., Al-Hazza'a, A., Karam, H., Al-Wadi, M., Al-Dhafeeri, A. & Al-Rowaih, A., Insights into the evaluation of the abiotic and biotic degradation rate of commercial pro-oxidant filled polyethylene (PE) thin films. *J. Environ. Manage.*, **250**, 109475, 2019.

[2] Parker, L., Fast facts about plastic pollution, 2018. https://news.nationalgeographic.com/2018/05/plastics-facts-infographics-ocean-pollution/.

[3] Boomerang Alliance, Marine plastic pollution by the numbers, 2016. https://www.boomerangalliance.org.au/sources.

[4] Colín-Chávez, C. et al., The release of carotenoids from a light-protected antioxidant active packaging designed to improve the stability of soybean oil. *Food Bioprocess Technol*, **7**, pp. 3504–3515, 2014.

[5] Al-Salem, S.M., Establishing an integrated databank for plastic manufacturers and converters in Kuwait. *Waste Manage.*, **29**, pp. 479–484, 2009.

[6] Tilstra, L. & Johnsonbaugh, D., The biodegradation of blends of polycaprolactone and polyethylene exposed to a defined consortium of fungi. *J Environ Polym Degr*, **1**, pp. 257–267, 1993.

[7] Guarás, M.P., Alvarez, V.A. & Ludueña, L.N., Processing and characterization of thermoplastic starch/polycaprolactone/compatibilizer ternary blends for packaging applications. *J Polym Res*, **22**, p. 165, 2015.

[8] Alariqi, S.A.S. & Singh, R.P., Effect of γ-Dose rate on biodegradation of γ-sterilized biomedical polyolefins. *J Polym Environ*, **18**, pp. 600–607, 2010.

[9] Al-Salem, S.M. et al., Study of the degradation profile for virgin linear low-density polyethylene (LLDPE) and polyolefin (PO) plastic waste blends. *J Mater Cycles Waste Manag*, **21**, pp. 1106–1122, 2019.

[10] Lee, J.A., Kontopoulou, M. & Parent, J.S., Time and shear dependent rheology of maleated polyethylene and its nanocomposites. *Polymer*, **45**, pp. 6595–6600, 2004.

[11] Wood-Adams, P.M., Dealy, J.M., DeGroot, A.W. & Redwine, O.D., Effect of molecular structure on the linear viscoelastic behavior of polyethylene. *Macromolecules*, **33**(20), pp. 7489–7499, 2000.

[12] Wood-Adams, P.M. & Dealy, J.M., Using rheological data to determine the branching level in metallocene polyethylenes. *Macromolecules*, **33**(20), pp. 7481–7488, 2000.

[13] Dumoulin, M.M., Utracki, L.A. & Carreau, P.J., Rheological properties of LLDPE/PP blends I. Dependence of tensile properties on composition. *Progress and Trends in Rheology II*, eds, H. Giesekus & M.F. Hibberd, Steinkopff: Heidelberg, 1988.

[14] Naddeo, C., Guadagno, L., De Luca, S., Vittoria, V. & Camino, G., Mechanical and transport properties of irradiated linear low density polyethylene (LLDPE). *Polym Degrad Stab*, **71**, pp. 239–247, 2001.

[15] Biswas, J. et al., Linear low density polyethylene (LLDPE)/zeolite microporous composite film. *Macromol. Res.*, **11**, pp. 357–367, 2003.

[16] Pang, M.M., Pun, M.Y. & Ishak, Z.A.M., Thermal, mechanical, and morphological characterization of biobased thermoplastic starch from agricultural waste/polypropylene blends. *Polym Eng Sci*, **54**, pp. 1357–1365, 2014.

[17] Osman, H. & Zakaria, M.H., Effects of durian seed flour on processing torque, tensile, thermal and biodegradation properties of polypropylene and high density polyethylene composites. *Polym Plastic Tech Eng*, **51**, pp. 243–250, 2012.

[18] Azhari, C.H. & Wong, S.F., Morphology-mechanical property relationship of polypropylene/starch blends. *Pakistan J Bio Sci*, **4**, pp. 693–695, 2001.

[19] Berruezo, M., Ludueña, L.N., Rodriguez, E. & Alvarez, V.A., Preparation and characterization of polystyrene/starch blends for packaging applications. *J Plastic Film Sheet*, **30**, 2014.

[20] Elanmugilan, M., Sreekumar, P.A., Singha, N.K., Al-Harthi, M.A. & De, S.K., Aging of low-density polyethylene in natural weather, underground soil aging and sea water: Effect of a starch-based prodegradant additive. *Polym Eng Sci*, **53**, pp. 2389–2397, 2013.

[21] Beg, M.D.H. & Pickering, K.L., Accelerated weathering of unbleached and bleached Kraft wood fibre reinforced polypropylene composites. *Polym Degrad Stab*, **93**, pp. 1939–1946, 2008.

[22] Vogt, N.B. & Kleppe, E.A., Oxo-biodegradable polyolefins show continued and increased thermal oxidative degradation after exposure to light. *Polym Degrad Stab*, **94**, pp. 659–663, 2009.

[23] Al-Salem, S.M., Abraham, G., Al-Qabandi, O.A. & Dashti, A.M., Investigating the effect of accelerated weathering on the mechanical and physical properties of high content plastic solid waste (PSW) blends with virgin linear low density polyethylene (LLDPE). *Polym Test*, **46**, pp. 116–121, 2015.

[24] Al-Salem, S.M., Al-Dousari, N.M., Joseph Abraham, G., Aromin D'Souza, M., Al-Qabandi O.A. & Al-Zakri W., Effect of die head temperature at compounding stage on the degradation of linear low density polyethylene/plastic film waste blends after accelerated weathering. *Int J Polymer Sci*, 5147209, 2016.

[25] La Mantia, F.P., Recycling of heterogeneous plastics wastes: I-Blends with low-density polyethylene. *Polym Degrad Stab*, **37**, pp. 145–148, 1992.

[26] Martínez-Romo, A., González-Mota, R., Soto-Bernal, J.J. & Rosales-Candelas, I., Investigating the degradability of HDPE, LDPE, PE-BIO, and PE-OXO films under UV-B radiation. *J Spec*, 586514, 2015.

[27] George, S.C. & Thomas, S., Transport phenomena through polymeric systems. *Prog Polym Sci*, **26**, pp. 985–1017, 2001.

[28] Ojeda, T.F.M. et al., Abiotic and biotic degradation of oxo-biodegradable polyethylenes. *Polym Degrad Stab*, **94**, 2009.

[29] Xingzhou, H., Wavelength sensitivity of photo-oxidation of polyethylene. *Polym Degrad Stab*, **55**, pp. 131–134, 1997.

[30] Gugumus, F., Contribution to the photolysis of hydroperoxides in polyethylene. *Polym Degrad Stab*, **27**, pp. 19–34, 1990.

[31] D'Amelia, R.P., Gentile, S., Nirode, W.F. & Huang L., Quantitative analysis of copolymers and blends of Polyvinyl Acetate (PVAc) using Fourier Transform Infrared Spectroscopy (FTIR) and Elemental Analysis (EA). *World J Chem Edu*, **4**, pp. 25–31, 2016.

[32] Nazareth, M., Marques, M.R.C., Leite, M.C.A. & Castro, Í.B., Commercial plastics claiming biodegradable status: Is this also accurate for marine environments? *J Hazard Mater*, **366**, pp. 714–722, 2019.

[33] Küpper, L., Gulmine, J.V., Janissek, P.R. & Heise, H.M., Attenuated total reflection infrared spectroscopy for micro-domain analysis of polyethylene samples after accelerated ageing within weathering chambers. *Vibrational Spec*, **34**, pp. 63–72, 2004.

[34] Nikolić, M.A.L et al., The challenges in lifetime prediction of oxodegradable polyolefin and biodegradable polymer films. *Polym Degrad Stab*, **145**, pp. 102–119, 2017.

[35] Hayes, D.G. et al., Effect of diverse weathering conditions on the physicochemical properties of biodegradable plastic mulches. *Polym Test*, **62**, pp. 454–467, **2017**.

[36] Portillo, F., Yashchuk, O. & Hermida, É., Evaluation of the rate of abiotic and biotic degradation of oxo-degradable polyethylene. *Polym Test*, **53**, pp. 58–69, 2016.

[37] Finzi-Quintão, C.M., Novack, K.M. & Bernardes-Silva, A.C., Identification of biodegradable and oxo-biodegradable plastic bags samples composition. Presented at *Macro Symposia*, 2016.

[38] Oliveira, T.A., Oliveira, R.R., Barbosa, R., Azevedo, J.B. & Alves, T.S., Effect of reprocessing cycles on the degradation of PP/PBAT-thermoplastic starch blends. *Carbohydrate Polym*, **168**, pp. 52–60, 2017.

[39] Moldovan, A., Paţachia Buican, R. & Tierean, M.H., Characterization of polyolefins wastes by FTIR spectroscopy. *Bullet Transilv*, **5**, 2012.

[40] Musioł, M., Rydz, J., Janeczek, H., Radeck, I., Jiang, G. & Kowalczuk, M., Forensic engineering of advanced polymeric materials Part IV: Case study of oxo-biodegradable polyethylene commercial bag – Aging in biotic and abiotic environment. *Waste Mange*, **64**, pp. 20–27, 2017.

[41] European Commission (EC), report from the commission to the European parliament and the council on the impact of the use of oxo-degradable plastic, including oxo-degradable plastic carrier bags, on the environment, 2018. http://ec.europa.eu/environment/circular-economy/pdf/oxo-plastics.pdf.

[42] Al-Salem, S.M., Influential parameters on natural weathering under harsh climatic conditions of mechanically recycled plastic film specimens. *J Environ Mange*, **230**, pp. 355–365, 2019.

[43] Al-Salem, S.M. et al., Thermal response and degressive reaction study of oxo-biodegradable plastic products exposed to various degradation media. *Int. J. Polym. Sci.*, 9612813, 2019.

[44] Al-Salem, S.M., Uddin, S. & Al-Yamani, F., An assessment of microplastics threat to the marine environment: A short review in context of the Arabian/Persian Gulf. *Marine Environ Res*, **159**, 104961, 2020.

[45] Al-Salem, S.M., Uddin, S. & Lyons, B., Evidence of microplastics (MP) in gut content of major consumed marine fish species in the State of Kuwait (of the Arabian/Persian Gulf). *Marine Poll Bullt*, **154**, 111052, 2020.

STUDY OF COMMERCIAL THERMOPLASTIC BIODEGRADABLE POLYESTER RESIN AS A SOLID WASTE MITIGATION ROUTE USING ASTM D 5988-18

MOHAMED WAHEED KISHK, HAJAR JAWAD KARAM, MASUMAH AL-QASSIMI,
AISHA ABDULLAH AL-ROWAIH, MAJED HAMEED AL-WADI & SULTAN MAJED AL-SALEM
Environment and Life Sciences Research Centre, Kuwait Institute for Scientific Research, Kuwait

ABSTRACT

Developing countries are faced with an overgrowing solid waste problem that various governmental bodies are trying to combat using either source mitigation or *end of pipe* solutions. Furthermore, the accumulation of plastic solid waste, in particular, is causing major concern to various entities around Gulf Council Countries (GCC), where various efforts are undertaken at state level to mitigate its accumulation from point sources. One such, is the development of standards for biodegradable plastics, namely comparing market available grades of oxo (pro-degradant enhanced) biodegradables against thermoplastic resin masterbatches. In this work, we have initiated such research efforts that takes into consideration governmental demands in the GCC and have considered a commercially available polyester masterbatch to study using the standard test method of the American Society of Testing and Materials D 5988-18 "Determining Aerobic Biodegradation of Plastic Materials in Soil". Laboratory controlled conditions were used to initiate the study of determining the degree and rate of aerobic biodegradation of the material in contact with soil that represents an active landfill site within the State of Kuwait. The masterbatch resin considered was previously claimed to be certified by the manufacturer in accordance with EN 17033 as a bio-based plastic derived from corn rendering the material as a modified polyester and polyactide. A potato starch (MSDS CAS No. 009005-84-9) purchased from Loba Chemie Company was used for positive control experimental runs. The soil was aerated with the compost in laboratory conditions and mixed in a 1:25 ratio (e.g. 18:450 g compost/g soil) with a moisture content of 80–100% maintained throughout the full duration of the experiment by adding distilled water. The biodegradation rate (%) was evaluated based on the carbon evolution estimation from the stoichiometry of the degradation reaction. The reported results can be used as a comparison baseline for both standards development (for market purposes) and applicability of material as a plastic waste mitigation technique for use.

Keywords: biodegradation, polyester, starch, ASTM D 5988, waste management.

1 INTRODUCTION

With the increasing threat of plastic solid waste (PSW) namely due to its non-degradable nature, water clogging of soil and micro/plastics fragments distribution in environmental sinks, efforts have increased the world over to overcome this issue [1]–[3]. One of the most promising solutions to the PSW accumulation problem is the development of biodegradable plastics. This class of polymers seem to represent themselves as an advantageous solution especially in developing world countries where PSW generation is considered to be higher than other parts of the world [1], [4]. In more specific terms, Arab Gulf Council Countries (GCC) have been trying to combat this problem for the past decade or so and various research efforts have taken place with successes in developing various standards and determining the applicability of such materials to their arid environments [5]–[8]. GCC countries are also characterised by being high waste generating countries, where for instance a relatively small nation such as Kuwait produces over 200 ktpa of PSW [3].

Biodegradability of plastics is also a matter that has been researched to a great extent in recent years [9]–[11]. There exist various standardised testing protocols that have been developed by various bodies such as the American Society of Testing and Materials (ASTM)

WIT Transactions on Ecology and the Environment, Vol 247, © 2020 WIT Press
www.witpress.com, ISSN 1743-3541 (on-line)
doi:10.2495/WM200111

and the International Standards Organization (ISO) to measure the extent of biodegradability for polymers. The majority of these tests involve soil burial, enzymatic degradation and carbon evolution measurements. For a review of such tests methods, readers are referred to Al-Salem et al. [5]. Nonetheless, it is of immense importance to have these tests mimic the reality of the region and the nature of the studied samples. To this end and following this research group's recent work in this research area [5]–[7], we have applied the standard test method of ASTM D 5988-18 "Determining Aerobic Biodegradation of Plastic Materials in Soil" on a number of biodegradable plastic blends that were prepared and compounded within our facilities. The plastics mimic those that are used on a daily basis within regional (west Asian) and international markets and have been popularized as of late namely on the Kuwaiti market. The test method has been also modified using soil that resembles landfill sites in Kuwait to determine the actual fate of such materials if they were discarded as a polymeric waste material. In addition, biodegradation has been enforced on real-life reclaimed PSW blended with pro-degradants to study the possibility of eliminating PSW using biodegradation as a waste minimization solution. To the best of our knowledge, such a study angle has not been attempted in past research efforts on such plastic blends.

2 MATERIALS AND METHODS

2.1 Materials

PSW originating from municipal sources (in films form) was outsourced from waste recyclers within Kuwait. The materials were previously characterised for type and percentile distribution as shown in [3]. Linear low density polyethylene (LLDPE) is the most common plastic film grade used the world over, and was used in this work as a 3 mm translucent pellets with a reported density of 0.918 g cm^{-3} and a melt flow index (MFI) of 2 g 10 min^{-1}. A PDQ-M Willow Ridge Plastics (Inc.) oxo-pro-degradant additive was used in the masterbatch preparation, which has a reported Tm of 110°C and was used as tan pellets. A BASF Ecovio certified compostable masterbatch (thermoplastic starch, TPS) declared as a finished product was used as white translucent pellets. The masterbatch is claimed to be certified by the manufacturer as a biobased plastic (modified polyester and polylactide) with a melting point of 120°C. Polycaprolactone (PCL) was also acquired and used as a reference biodegradable plastic material (Polysciences/CAT 19561).

2.2 Plastic film compounding

PSW was air blown with an air jet (2 bar) and were milled to 3 mm flakes using an industrial Tecnova S.R.L. cutting mill and pelletized to 3 mm pellets using a Tecnova single screw extruder (L/D = 30) at 40 bars pressure and 180 rpm. The extrusion zone temperatures were maintained between 155 and 195°C at die. Single screw extrusion (SSE) was performed using a Dolci extruder (L/D ratio 1/28) using 10 kg as masterbatch mix ratios maintained between 165 and 200°C for extrusion zones. A flow rate of 50 kg h^{-1} was kept throughout with an operating pressure of 6 bars. The study samples were compounded with the following dry weight percentages (wt./wt.%) to mimic the grades available on the market including comparative assessment with virgin LLDPE and a 100 (wt.%) PSW film: LLDPE (100 wt.%), PSW (100 wt.%), LLDPE/PDQ-M (97/3 wt.%/wt.%), PSW/PDQ-M (97/3 wt.%/wt.%), LLDPE/TPS (97/3 wt.%/wt.%) and PSW/TPS (97/3 wt.%/wt.%). TPS and PCL were used as received pellets as described below. All samples were extruded as sheets of

100 µm thickness meeting standard market specifications and as an average of major film applications [12].

2.3 Biodegradation test

Biodegradation was performed based on ASTM D5988-18 [13] to evaluate the degree and rate of aerobic biodegradation of plastic materials relative to reference material in contact with soil under laboratory controlled conditions.

2.3.1 Soil source and characterization

A mixture of natural soil and mature compost was used as a test matrix. Natural surface soil samples (50 kg) was obtained locally originating from Maqwa'a area mimicking waste landfill soil [5]. The soil was used fresh (as received) to ensure presence of active microbiota after sieving it to > 2 mm particle size which allows the components to be well distributed. Soil was mixed local compost (sheep manure) in a 1:25 ratio. The pH value (in a 20 to 20 (w/v) slurry), water content (after drying for 16 h at 110 ± 5°C), ash content (by igniting the oven-dried sample in a furnace at 440 ± 40°C), organic matter content, water holding capacity (WHC), total carbon and nitrogen, the microbial population of the soil were measured according to standard methods of ASTM D4972-19, ASTM D2974-14, BS EN ISO 846, Dumas dynamic method and Gupta method, respectively [14]–[19]. The pH value was 7.8; water content was 3.7%, ash content was 2.5% and the organic matter was 40.2%.

2.3.2 Experimental set-up

The test was performed in triplicate for each of the technical control (desiccator that contains only the absorbing solution in a sealed air-tight system), soil blank (showing the background activity of the soil), positive reference material (to check the viability of the soil microbial community), and test materials (triplicate for each test material). Starch powder from Loba Chemie Company were employed as a reference material for the biodegradation test in the soil. The soil was weighed and mixed with compost in a 1:25 ratio (18:450 g compost/g soil) in a separate container to get a homogeneous mixture (test matrix). Thereafter, the soil was transferred to the bottom of each vessel (150 mm desiccators with 4 litres internal volume equipped with an air-tight seal). The final soil water content was adjusted to 12.7% (about 50% of the water holding capacity). Approximately 600 mg of test material (sufficient quantity to yield enough carbon dioxide that allows for better monitoring of the actual biodegradation rate for the test materials), in the form of pellets or films that were cut out square-shaped coupons with a dimension of approximately 2 cm^2. The mass was recorded for each sample. To improve the contact of the tested material with the microbiota, the powder/pellets were mixed thoroughly into the soil. In the case of film coupons, the tested material was evenly spread and pressed with a spatula in the soil. Each vessel contained two 100 mL beakers, one with 20 ml of potassium hydroxide (0.5 N KOH) and the other containing 50 ml of distilled water and were placed on the perforated plate inside the vessel.

2.3.3 CO_2 analysis

The CO_2 analysis was measured based on the CO_2 evolved in each vessel by analysing its reaction with KOH to produce potassium carbonate (K_2CO_3) and unreacted KOH. The titration was performed with hydrochloric acid (0.5 N HCl) to a phenolphthalein end-point according to the following equations:

$$2\,KOH + CO_2 \rightarrow K_2CO_3 + H_2O, \tag{1}$$

$$KOH + HCl \rightarrow KCl + H2O \text{ (at pH equals to 7)}, \quad (2)$$

$$K_2CO_3 + HCl \rightarrow KHCO_3 + KCl \text{ (at pH equals to 8.5)}. \quad (3)$$

The amount of CO_2 produced by a test material was calculated by the difference (in ml of titrant) between the test and blank vessels. The volume of HCl consumed in ml, was used to calculate the molar concentration (mmol) of CO_2. The readers are referred Al-Salem et al. [5] for further methodological details; and Fig. 1 for a pictorial depiction of the test set-up.

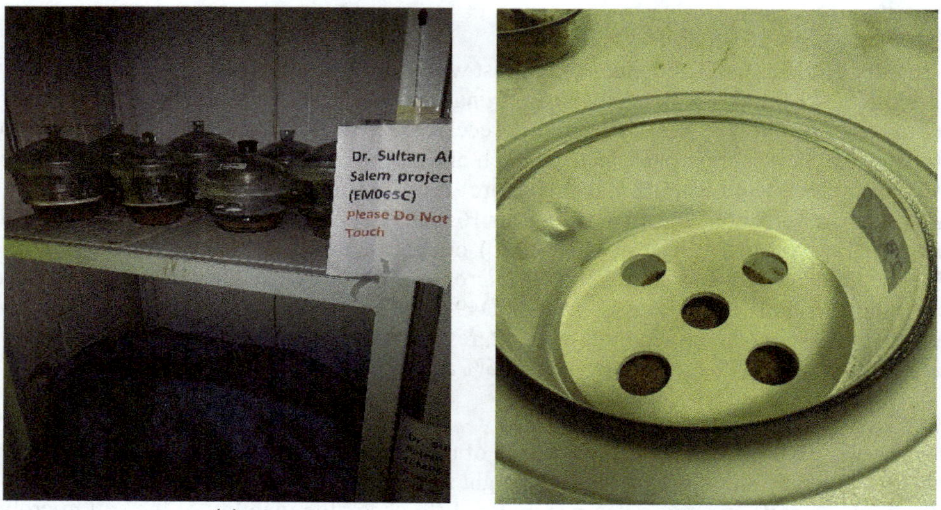

(a) (b)

Figure 1: (a) Dark room set-up for biodegradation test; and (b) Open glass desiccator (4 litres) used in biodegradation tests.

3 RESULTS AND DISCUSSIONS

To be able to determine the extent of biodegradation for the materials used, various parameters must be observed within the experimental matrix. These will include the rate of biodegradation, the nature and experimental error of the results and the carbon evolution with respect to the study time. The total experimental time for the test method is given at a maximum by the standard protocol test methods followed as six months in total. Fresh soil was used to assure that the soil microbiota is of maximum diversity and of high activity in a suitable degrading environment. The initial pH value for the soil used was found to be 7.8. This indicates that the media if of a weak alkali nature and has the potential to contain a typical microbial population [13]. The total organic content of the soil was estimated at 40.2% which indicates that it is of a suitable ratio to maintain microbial growth. The water content was adjusted to about 50% of the maximum WHC and was controlled throughout the experimental period as the water availability is a crucial parameter for the microbial growth and metabolic activity and having a significant effect on the biodegradation processes [20], [21]. The total number of microorganisms (including bacteria and fungi) in the soil was 4×10^6 cfu per gram of the test matrix (estimated by determining average colony forming units, cfu, using spread plate method). ASTM D5988 was used to assess the degree of aerobic biodegradation on the compound films in soil by measuring the evolved CO_2 as a function of exposure time to determine the extent of their biodegradability, as it is one of the latest and approved standard methods. Based on previous work [13] and the recommendation of the ASTM, a KOH trap was used as an alternative to barium hydroxide solution ($Ba(OH)_2$) to facilitate the diffusion of CO_2 into the absorbing medium. The biodegradation was calculated from the ratio between the net CO_2 production and the theoretical CO_2 based on the carbon content. The plastic material was tested in the form of pellets and film as it is acceptable for test specimens to be introduced in the form of films, pieces, fragments, powders, or shaped articles [13]. Biodegradation extent is shown in Fig. 2. Each material studied is represented with respect to real-experimental time (in days) stemming from the mean value of three triplicate runs.

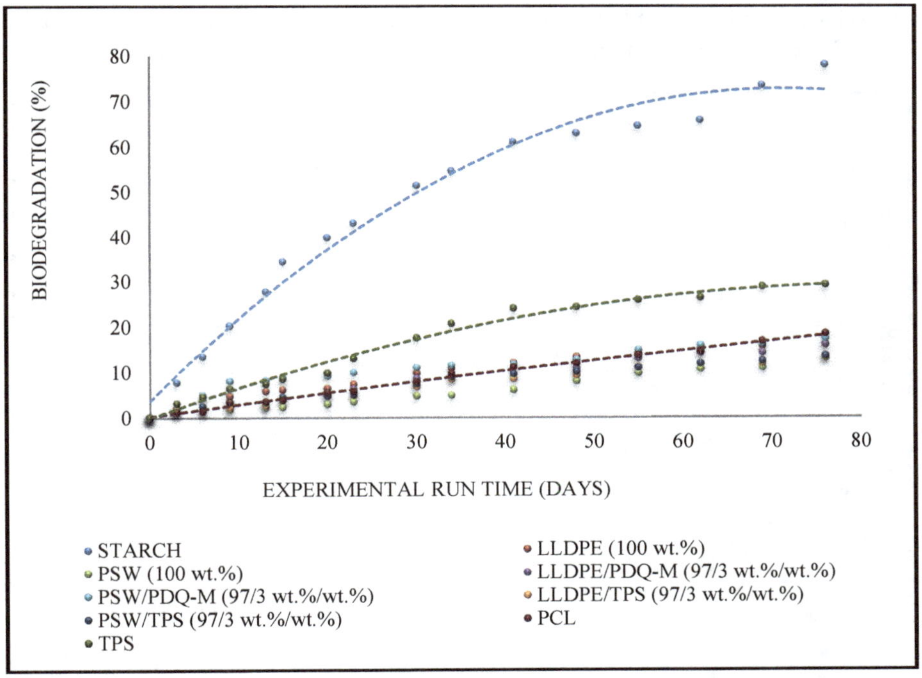

Figure 2: Biodegradation rate (%) estimated (to date) for 76 days of experimental run-time.

The starch reference material was the highest in biodegradation rate as anticipated at the start of the experimental run. The material exhibited over 50% of degradation rate in less than 50 days of continuous monitoring. The starch material also showed a percentage of biodegradability of > 70% after 70 days which meets the validation criteria for the test method [13]. This on the other hand indicates that the soil was active and the test are valid. Furthermore, the TPS material was the second in-line, in terms of biodegradation rate (Fig. 2). The mineralization for TPS reached 29 ± 0.28% in 76 days, rendering it second highest in degradation rate to the starch (78 ± 0.44%). PCL, the biodegradable polymer used as well for comparative reasons, was third in the depicted results showing a rate of biodegradation of some 18.3 ± 0.22% by the end of the testing duration which is noted to be a delayed response considering the cycle for this materials nature. This also indicates that the soil conditions aforementioned might not be suitable for such materials under certain arid-land conditions. The humidity levels are also suspected to be a crucial element for the degradation of PCL under certain conditions. Šerá et al. [22] showed that the temperature of the incubation temperature from 25 to 37°C, increases the biodegradation rate of certain biodegradable plastics. At this stage, it should be noted that ASTM D 5988 is somewhat equivalent to the standard test protocol of ISO 17556. Mater-Bi HF03V1, a commercially available TPS made with similar characteristics to the TPS used in this work was studied using similar test conditions by Pischedda et al. [23]. At 15 and 20°C, the mineralisation (biodegradation rate) was estimated at some 20% after 100 days which is considerably lower than the estimated rate in this work. The behaviour observed in the aforementioned curve could also be explained by the rapid increase in microbial population at the beginning of the test period are readily available for assimilation which get depleted with experimental time [24]. Fig. 3 shows the degradation rate of the commercial plastics considered in this work.

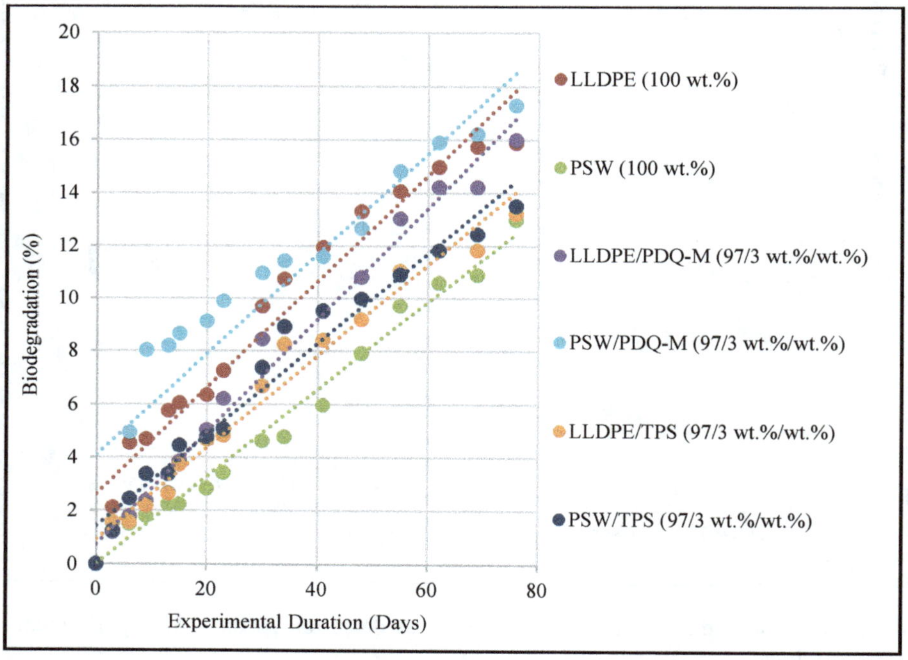

Figure 3: Biodegradation rate (%) estimated for the experimental run-time showing commercial polyolefin plastics considered.

The least favourable material in terms of biodegradation rate after 76 days of testing, was the PSW on its own used in the form of films simulating discarded waste buried in landfill sites. The material showed a 12.9% rate of biodegradation which was lower than the virgin LLDPE used in this work (15.8%). This could be attributed to the fact that the PSW was exposed to more heating cycles/loops that increases its crystallinity and hardens its structure compared to the virgin LLDPE. On the other hand, initial testing trials of the concept of inducing the biodegradation rate of PSW proved viable. This is supported by the fact that the PSW filled with pro-degradant (by 3 wt.%) was degrading faster than the other materials considered in this study. The biodegradation rate of the PSW/PDQM (97/3%), LLDPE/PDQM (97/3%), and LLDPE (100%) films was 17.3 ± 0.33%, 16 ± 0.12% and 15.89±0.23%, respectively.

4 CONCLUSION

In the context of environmental management, it is almost a certainty that plastic solid waste accumulation and generation, is favoured and serious mitigation measures should be taken to combat it. In this work, plastic and virgin commercial grade plastics were tested after studying their biodegradability with both a pro-degradant additive and a thermoplastic starch material. The materials were also compared to biodegradable polymers and resins. The test method of ASTM D 5988-18 was used to study the biodegradation using a soil matrix that simulates landfill environments in arid-lands. Rigorous testing of the soil material was also conducted to have a clear overview of the testing environment. In terms of the materials tested (to date), the pro-degradant additive that rendered both the virgin linear low density polyethylene and plastic solid waste, made it more susceptible to biodegradation and fragmentation. This is when compared to the other tested material in this work excluding the naturally (as is) biodegradables. This indicates that it is possible to consider the mechanical recycling of plastic waste with the addition of pro-degradants to make them more favourable to biodegradation. There is a certain concern when it comes to interrupting these preliminary results obtained in this work. The main barrier is to assess the eco-friendly design and aspect of having metal added pro-degradants that might increase the toxicity of the plastics when exposed to high humidity environments. Therefore, and since the materials tested in this work with starch based components show promising results, it might be worth in the future to consider a comparative assessment of their life cycle or impact assessment works.

ACKNOWLEDGEMENT

The authors are grateful to the Kuwait Institute for Scientific Research (KISR) for funding research project EM065K.

REFERENCES

[1] Yang, H.S., Yoon, J.S. & Kim, M.N., Dependence of biodegradability of plastics in compost on the shape of specimens. *Polymer Degradation and Stability*, **87**(1), pp. 131–135, 2005.

[2] Al-Salem, S.M., Uddin, S. & Lyons, B., Evidence of microplastics (MP) in gut content of major consumed marine fish species in the State of Kuwait (of the Arabian/Persian Gulf). *Marine Pollution Bulletin*, 2020.

[3] Al-Salem, S.M., Abraham, G., Al-Qabandi, O.A. & Dashti, A.M., Investigating the effect of accelerated weathering on the mechanical and physical properties of high content plastic solid waste (PSW) blends with virgin linear low density polyethylene (LLDPE). *Polymer Testing*, **46**, pp. 116–121, 2015.

[4] Kaza, S., Yao, L., Bhada-Tata, P. & Van Woerden, F., *What a Waste: A Global Snapshot of Solid Waste Management to 2050*, World Bank: Washington, DC. DOI: 10.1596/978-1-4648-1329-0, 2018.

[5] Al-Salem, S.M., Sultan, H.H., Karam, H.J. & Al-Dhafeeri, A.T., Determination of biodegradation rate of commercial oxo-biodegradable polyethylene film products using ASTM D 5988. *Journal of Polymer Research*, **26**(7), p. 157, 2019.

[6] Al-Salem, S.M., Al-Hazza'a, A., Karam, H.J., Al-Wadi, M.H., Al-Dhafeeri, A.T. & Al-Rowaih, A.A., Insights into the evaluation of the abiotic and biotic degradation rate of commercial pro-oxidant filled polyethylene (PE) thin films. *Journal of Environmental Management*, **250**, 109475, 2019.

[7] Al-Salem, S.M. et al., Thermal response and degressive reaction study of oxo-biodegradable plastic products exposed to various degradation media. *International Journal of Polymer Science*, 9612813, 2019.

[8] Abdallah, M., Shanableh, A., Shabib, A. & Adghim, M., Financial feasibility of waste to energy strategies in the United Arab Emirates. *Waste Management*, **82**, pp. 207–219, 2018.

[9] Castro-Aguirre, E., Auras, R., Selke, S., Rubino, M. & Marsh, T., Insights on the aerobic biodegradation of polymers by analysis of evolved carbon dioxide in simulated composting conditions. *Polymer Degradation and Stability*, **1**(137), pp. 251–271, 2017.

[10] Oberbeckmann, S., Loeder, M.G., Gerdts, G. & Osborn, A.M., Spatial and seasonal variation in diversity and structure of microbial biofilms on marine plastics in Northern European waters. *FEMS Microbiology Ecology*, **90**(2), pp. 478–492, 2014.

[11] Großkopf T. & Soyer O.S., Synthetic microbial communities. *Current Opinion in Microbiology*, **18**, pp. 72–77, 2014.

[12] Al-Salem, S.M., Influential parameters on natural weathering under harsh climatic conditions of mechanically recycled plastic film specimens. *Journal of Environmental Management*, **230**, pp. 355–365, 2019.

[13] ASTM International, ASTM D5988-18, Standard test method for determining aerobic biodegradation of plastic materials in soil. ASTM International: West Conshohocken, PA, 2018. www.astm.org/Standards/D5988.html. Accessed on: 22 Nov. 2019.

[14] ASTM International, D4972-19, Standard test methods for pH of soils. ASTM International: West Conshohocken, PA, 2019. DOI: https://doi.org/10.1520/D4972-19.

[15] ASTM International, D2974-14, Standard test methods for moisture, ash, and organic matter of peat and other organic soils. ASTM International: West Conshohocken, PA, 2014. DOI: https://doi.org/10.1520/D2974-14.

[16] British Standards (BS), ISO 846, Implementation of the International Standards Organization, Plastics: Evaluation of the action of microorganisms, 2019. https://doi.org/10.3403/30367618.

[17] Bremner, J.M., Nitrogen-total. *Methods of Soil Analysis: Part 3 Chemical Methods*, eds D.L. Sparks, A.L. Page, P.A. Helmke, R.H. Loeppert, P.N. Soltanpour, M.A. Tabatabai, C.T. Johnston & M.E. Sumner, Soil Science Society of America and American Society of Agronomy: Madison, WI, pp. 1085–1121, 1996.

[18] Krotz, L. & Giazzi, G., Elemental analysis: CHNS/O characterization of polymers and plastics, Application Note 42230, Thermo Fisher, 2016. https://assets.thermofisher.com/TFS-Assets/CMD/Application-Notes/an-42230-oea-chnso-polymers-plastics-an42230-en.pdf.

[19] BS 7755-3.8, Soil quality. Chemical methods. Determination of organic and total carbon after dry combustion (elementary analysis), 1995. DOI: https://doi.org/10.3403/00577903.

[20] Grima, S., Bellon-Maurel, V., Feuilloley, P. & Silvestre, F., Aerobic biodegradation of polymers in solid-state conditions: A review of environmental and physicochemical parameter settings in laboratory simulations. *Journal of Polymers and the Environment*. **8**(4), pp. 183–195, 2000.

[21] Briassoulis, D. & Dejean, C., Critical review of norms and standards for biodegradable agricultural plastics, Part I: Biodegradation in soil. *Journal of Polymers and the Environment*, **18**(3), pp. 384–400, 2010.

[22] Šerá, J., Serbruyns, L., De Wilde, B. & Koutný, M., Accelerated biodegradation testing of slowly degradable polyesters in soil. *Polymer Degradation and Stability*, **171**, 109031, 2020.

[23] Pischedda, A., Tosin, M. & Degli-Innocenti, F., Biodegradation of plastics in soil: The effect of temperature. *Polymer Degradation and Stability*, **170**, 109017, 2019.

[24] Castro-Aguirre, E., Auras, R., Selke, S., Rubino, M. & Marsh, T., Insights on the aerobic biodegradation of polymers by analysis of evolved carbon dioxide in simulated composting conditions. *Polymer Degradation and Stability*, **137**, pp. 251–271, 2017.

POTENTIAL INTEGRATED SMART WASTE SEGREGATION FOR ALL STAKEHOLDERS

BARMAK SADRI[1], BARAT ALI KHAKPOOR[2] & MOHAMMAD AJZA SHOKOUHI[2]
[1]Faculty of Economics and Administrative Sciences, Ferdowsi University of Mashhad, Iran
[2]Department of Geography, Ferdowsi University of Mashhad, Iran

ABSTRACT

There are formal and informal recycling workers as a part of the mainstream and resource management system that are handling separation of waste at source duties instead of urban dwellers who are facing the loss of aesthetic value of their cities. Less recycling wastes remain in the strewed waste bins to serve a circular economy from this disgusting perspective where uncontrolled waste collection, recycling, treatment, and disposal is likely to have health impacts on such workers. Further, their operations cause dispersal of contamination and debris that do not sound effective and environmentally waste management for a clean city and a pleasant and healthy living environment that is attractive to residents and visitors. The increasing number of informal workers demonstrates unsuccessful strategies of separation of waste at source and aborted due to lack of citizens engagement. Recently smart technologies emerged that sound useful to tackle the issues due to its capabilities of monitoring waste stream. Hence, the investigated performance indicators by scholars to improve waste management, reviewed in this study to develop a model for smart waste recycling. The service design approaches employed to be compatible to local infrastructures of the studied cities of Tehran and Mashhad in Iran that the potential smart waste segregation model figured out, based on integrated smart identification of all stakeholders, to apply legislation for citizens, communities and advocates, and private sector, and to provide monitoring and measurement tools about recycling waste and 4R (reduction, reuse, recycling, recovery) procedures. The system could be implemented in similar metropolises like Ankara, who face such issues. Further local analysis for technology acceptance evaluation should apply to refer client types, generation, and local infrastructures whether a combination of different identification and pickup methods will meet their satisfaction and improve the waste management system.
Keywords: informal recycling workers, separation of waste at source, monitoring waste stream, integrated smart waste recycling, smart waste identification, potential smart waste segregation system, disgusting perspective of cities, SAAS, DSS, circular economy.

1 INTRODUCTION

The world's urban population has grown rapidly from 751 million to 4.2 billion in 2018 [1]. It expects to increase by 25% by 2050, Leading to a global population of 9.8 billion [2]. Now more than ever, cities are hot spots that threaten global ecological boundaries. Hence global sustainability pivots mainly on cities and that meeting this challenge requires an urgent change in urban policymaking and planning (Olazable and Pascual [3]). Therefore, we are facing lots of issues in our daily life in the cities as urban dwellers. The increasing volume and complexity of waste associated with the modern economy is posing a serious risk to ecosystems and human health [4]. Adequate and affordable solid waste collection services sound accessible in metropolises of Iran for citizens, where have been facing numerous informal waste workers in the recent decade due to failing separation at source strategies. There are remaining recycling waste management kiosks from previous management that still provide barter products instead of transferring all their clients' waste to the informal workers where they are accessible around every single waste bin.

A practical and environmentally waste management where contributes to a clean city, and a pleasant and healthy living environment needs urgent collaboration of both citizens and management to tackle informal workers issue. Fortunately, expanding digital technologies

WIT Transactions on Ecology and the Environment, Vol 247,
www.witpress.com, ISSN 1743-3541 (on-line)
doi:10.2495/WM200121

introduce new signs to struggle with complicated challenges like separation of waste at source during a scientific journey to figure out a potential smart waste segregation system that would be based on any further complementary research. What are the indicators of a potential smart waste segregation system for Tehran and Mashhad's target cities to engage all citizens that would be compatible with available infrastructure?

The aim of the research addresses the objective of minimizing the informal workers by maximizing the participation of citizens in the separation of waste at source, or less informal supply concludes less informal demand.

There is an urgent need to look into the issues of SWM and improve city leaders' ability to manage with shrinking budgetary support. Performance indicators (PIs) are measurement tools used by organizations to evaluate the success or failure of a given activity. He conducted a holistic review of articles to develop performance indicators for municipal solid waste management after analyzing 387 research papers published in the area of solid waste management (SWM) (Sanjeevi and Shahabudeen [5]). The initial SWM models were optimization models, and most dealt with minimizing costs (Berger et al. [6]). There were developments of new models by researchers as [7]–[10], with increasing complexity in SWM in the cities of the developing world, selection, or setting up of an optimum SWM system becomes difficult for technical and operation research professionals. Researches led to the use of various mathematical models and systems analysis techniques to develop integrated solid wastes management systems. By focusing on the application of general management techniques, the research papers grouped under various types in a flowchart makes it easier to understand the nature of SWM systems .Also, the authors grouped the papers for each decade under the 18 different types, such as cost reduction, optimization, reports/guidelines, and legislation. Then modern productivity tools like research focusing on the introduction of GIS/GPS, RFID, barcodes, and their implications in improving SWM via DSS, which started to appear only since 2000, synchronized with the commercial arrival of these technologies.

Further inferences from flowchart for the field research are characteristic of SWM issues and challenges in various cities, Models based on real-life case studies, current standards of SWM, and life cycle assessment. These would be researched this study by evaluating for recycling waste management. The challenge is separation at source by citizens to struggle with informal worker increment in Tehran and Mashhad. Our model, based on real-life case studies, compares the current facilities of recycling waste management services (consists of formal recycling workers by vans and recycling kiosks). To find out covert elements of the smart system, whether it is service or decision support system (DSS). A service-based smart system sounds more compatible with an integrated system by software as a service or SAAS system.

Armijo et al. [11] incorporated vital requirements such as social participation, social perception, and communication levels like those of the PIMS for the first time. They designed a driving force-pressure-state-impact response (DPSIR) model, involving 18 indicators like coverage, generation, composition, and efficiency. They also suggested that a user-friendly tool, which captured the complexity of the SWM, be developed. The performance criteria detailed by indicators such as the percentage of citizens in favour of recycling and those complying with the government regulations might not require from the PI's point of view. Nevertheless, in the interest of this study, a user-friendly tool considers generation and composition or different types of potential smart waste segregation system customers in favour of recycling waste. Those should comply with government regulations. As demonstrated by Sanjeevi and Shahabudeen [5] the study with 5PIs includes social perception and social participation (only 0.4%), which complies with the objectives of this study further suggested future research agenda for the future like community engagement where could have an important role in the improvement of recycling waste management, Maryono and Hasmantika [12] defined the high-level application of smart waste recycling with excellent community support. The provision of this application service must balance with socialization from the Government to the community. Hence, the mobile application facilitates communication between management and stakeholders in the era of digitization.as our platform is an integrated system. Therefore, the community acts should conduct into the system, and co-relation between citizens and communities would be possible to avoid any interruption in the data gathering process. Official data for MSW recycling often come from municipal governments, which in many developing countries focus on managing the MSW they collect (or which is collected on their behalf by the formal sector, leaving the collection of materials for recycling often to the informal sector). Official data, either at the city level or compiled from city-data by national governments, are thus likely to be under-reporting recycling rates (Wilson et al. [13]). The informal recycling that is often dominant in many developing countries is generally from mixed MSW. However, there can be a significant contribution from itinerant waste buyers who collect and pay for source-separated materials accumulated by householders or domestic servants. Community initiatives may also collect source-separated materials to gain funds for local charities (Wilson et al. [13]). Hence, the integrated smart system could be more useful for raising funds for local charities. In the worst case, 15% of citizens recycled waste in a month, which empowers charities to support the monthly wage of 5,000 informal workers according to the study despite waste management revenue from recycling waste. Segregation of MSW at source is critical to ensure that the waste separated into organic and dry recyclable fractions. Segregation of MSW at source, separating organic and dry recyclable fractions, is critical to avoid cross-contamination and to maintain the quality of the materials, which will lead to more effective recycling and divert waste from landfills. Further, segregated waste reduces health and safety-related risks to waste pickers and to the ecosystems around the waste treatment and disposal sites (Wilson et al. [13]). The issue addresses guidelines importance within smart system growth to lead better participation updated materials and illustration could affect the results. Economic instruments serve to steer stakeholders' behaviours and practices towards strategic goals through market-based incentives and disincentives. For example, a pay-as-you-throw (PAYT) is a charging system for residual (mixed) waste. It rewards people for segregating their waste, taxes on landfilling or incineration will discourage opting for these methods; fiscal benefits will encourage private companies' investment in SWM (Wilson et al. [13]). Economic instruments, as the legislation drivers are desirable in the recycling waste system. The incentives indicator for delivering waste to any verified picker, even charities versus disincentives, charge more tax for citizens if they do not deliver recycled waste at least once a month. Refer to the introduction elements of the potential smart waste segregation figure out as a smart recycling waste model shown in Fig. 1.

The connectivity of the elements as the conclusion of the research method covered by several iterations is the indicator of Smart Identification.

2 METHODS

To find out the expected specifications of a potential smart waste segregation system for target cities of Tehran and Mashhad. By considering discussion that the system would use as a service or SAAS. The appropriate approach to propose a prototype of a system is the service design approach that exploring the qualitative study. So inductive button-up research implied via a journey map of the autoethnographic approach. "Real" (i.e., rather academic) autoethnographic research might involve researchers immersing themselves for months

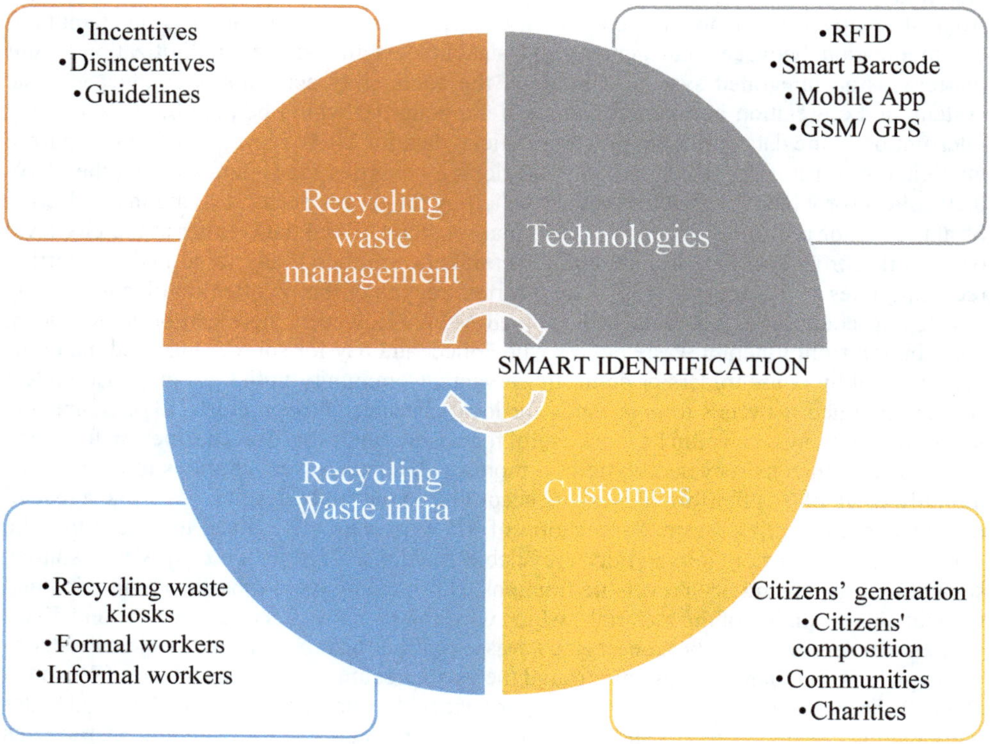

Figure 1: Smart waste stream identification.

within an organization. In service design, we often use a short version of this: team members explore a particular experience in the real situational context, mostly as customers or as employees. Autoethnography is often one of the first research methods undertaken as it helps researchers to interpret behaviours they will see when they observe participants. Also, it helps researchers to conduct interviews more efficiently and comprehensively when they already have a rough understanding of the subject matter. Hence, in this study, the author explored a particular experience of it self-document this using field notes, photographs, and both covert and overt interviews depend on conditions to avoid observer effect (Stickdorn et al. [14]).

2.1 Material

To analyze waste collection services of target cities, a simple persona between different generation and composition selected as a single middle age customer where produces an average amount of recycling waste per person. The journey of how it can deal with recycling waste describes the interrelation of recycling waste picker with the customer, and how waste pickers damage the aesthetic perspective of daily urban life despite dispersal of contamination and debris within their operation.

2.1.1 Persona

The persona case (Fig. 2) faces with recycling workers around the bins, sometimes he delivers recycling waste to the recycling waste kiosk and got other products instead of the value of the waste or leave the waste cost to the kiosk, sometimes he delivers waste to the informal workers. Waste management collection service runs twice a day appropriately. The customer has no problem with waste collecting service, and He regularly throws the waste to the bins.

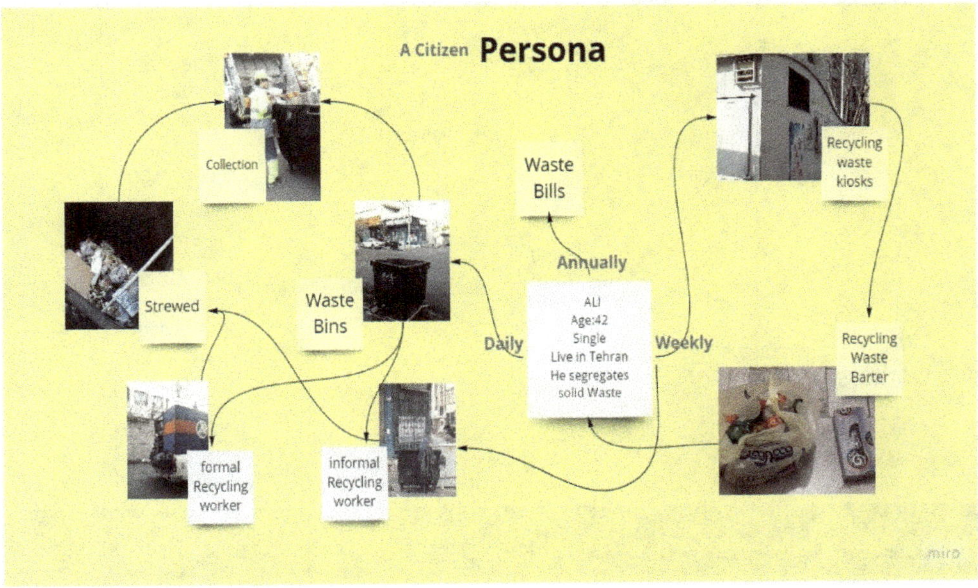

Figure 2: Citizen persona of waste tasks.

The waste management system has two stakeholders acting as recycling waste pickers, kiosks, and formal workers with pick-up where there is an allegation that some of the informal workers transfer their collected recycling waste to the formal workers. The issue is a conflict with other stakeholders as waste pickers who strew recycling waste from the bins cause an ugly image of the city.

2.1.2 Retrospective challenges

A retrospective analysis of what made a citizen MAD, SAD, and Glad mapped in Fig. 3 would bring some insights for understanding customer challenges. It segregates recycling waste, then delivers the recycling waste to the kiosk by car and must pick up barter goods instead of where the kiosk does not care about citizenship. The recycling waste workers made a disgusting perspective when strewing the bins for solid wastes. Their operations cause dispersal of debris and contamination due to mismanagement of recycling waste management.

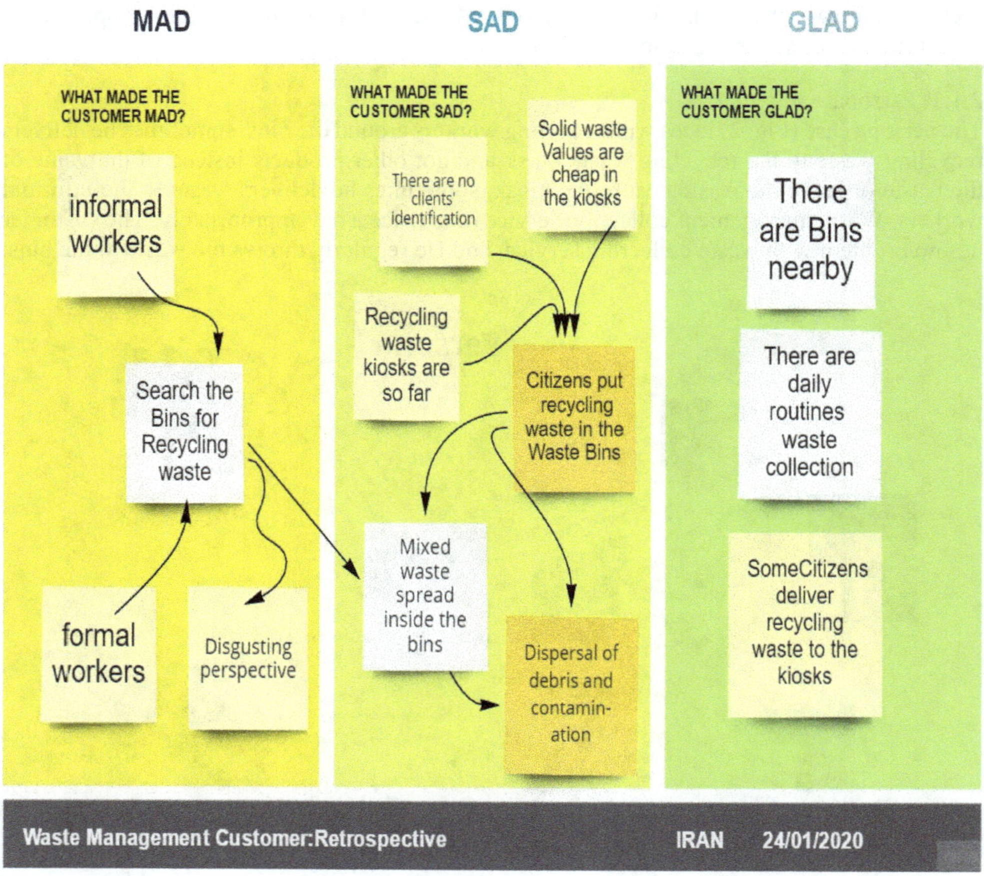

Figure 3: Customer retrospective.

2.2 Pre-ideation

Retrospective challenges develop some questions about the origin of the issue, whether those could lead us to find out a solution to tackle these challenges. Why have recycling workers expanded in the city? Due to the importance of the circular economy, economic conditions, illegal immigration, low participation of citizens at the separation of waste at source caused the abundant supply of recycling waste for workers. So, why do not more citizens participate in the separation of waste at source? Due to far facilities, Lack of obligation and no cost-benefit motivation, and why do some citizens participate in the separation of waste at source? Due to cultural reasons, environmental worries, the life cycle of materials, charity intention, and saving the planet for the next generation.

Hence, the above concludes a few hypotheses as if a system could decrease the supply of recycling waste at the bins, then recycling waste working demand would decline if citizens' engagement increases at the separation of waste at source, the supply of recycling waste would decrease. Finally, appropriate obligations and facilities raise citizens' participation.

2.2.1 Case studies from overseas

What kind of recycling waste collection systems has implemented to overcome the issues?

Case studies from overseas, from similar persona customer, from countries with less informal workers as demonstrated in Fig. 4 indicates that disincentives and guidelines incorporated as obligations by fines and cautions. It is plausible for waste management to apply obligations through the identification of customers by codes of the bins and RFID. Thus, identification addresses the system's key indicator.

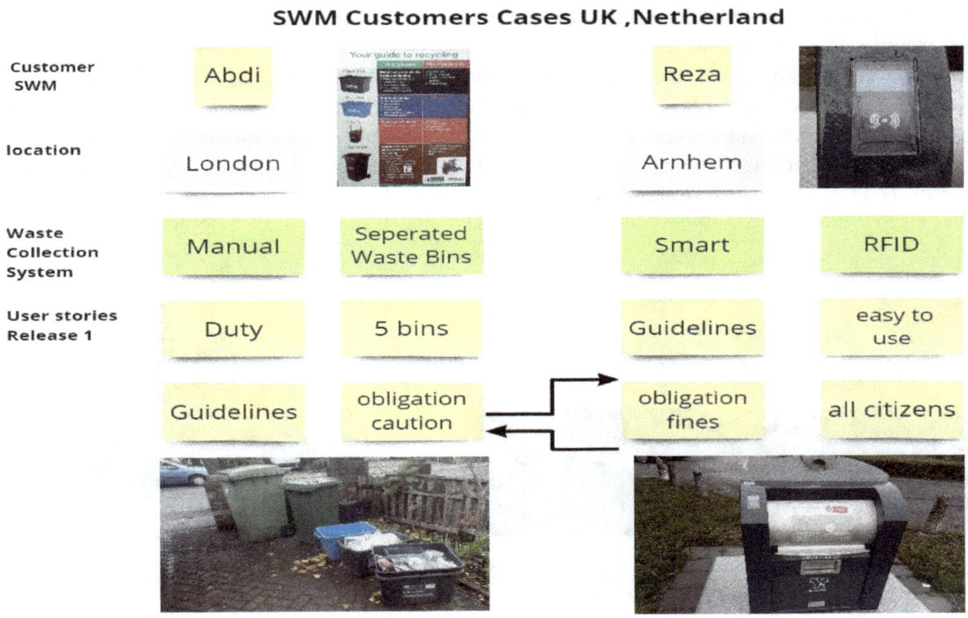

Figure 4: SWM Customers persona cases UK and The Netherlands.

Other criteria noticed by the Smart systems, although the further study of peer-reviewed articles, generally focused on smart waste collection systems by using different integrated methods for optimizing waste collection by container identification (Fig. 5) consisting of the control centre and bins sensors, IoT of bins, mobile application and more. This optimization applies to formal recycling workers with puck-up where should receive signals by customers instead of a container, however,

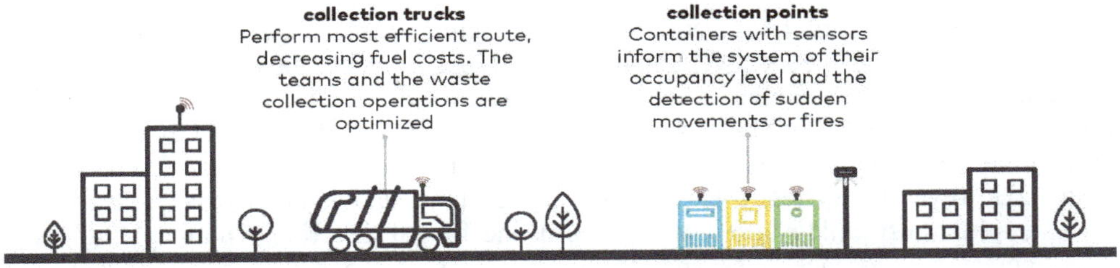

Figure 5: Smart waste collection system [15].

The technology case study demonstrates other features of smart systems usually used by the smart underground system; those identify customers by RFID, barcode bags, mobile applications and could measure and calculate incentives and disincentives. There are separate recycling waste inlets that customers throw their waste according to the guidelines. The underground network of tubes shown in Fig. 6 (left) is suitable for condense spatial parts of the cities where collection trucks could not drive in the alleys but needs costly infrastructure construction. The next one, Fig. 6 (right), digs the underground bins. It could be more compatible with the collection system consist of trucks and bins, even though it is still costly infrastructural operation for waste management, whether the cost of such systems could be affordable. Hence, the most critical issue of these systems is their cost of implementation. Despite the capability of the underground system to tackle objectives of this study like aesthetic perspective damaged by informal and formal recycling workers further avoiding dispersal of contamination and debris due to covered and unreachable bins and container as a holistic approach for separation at source and solid waste management system, but urgent action is a practical solution compatible to the current facilities.

Figure 6: Smart waste bins solutions [16], [17].

3 RESULTS AND DISCUSSION

3.1 Prototyping potential smart waste segregation system

According to a comprehensive review during the journey map, adequate information reached to propose a prototype that connects elements of the smart recycling waste model as a service blueprint to adopt local facilities and gather offers. The smart improvement in local recycling waste services based on smart identification facilitates separation at source operation to struggle with current issues and concludes the objectives of the study. The cross interrelation of stakeholders addresses collaboration between recycling waste management and customers. The generation of customers covered by different types of identification makes it plausible for elder or youth generation to connect to the system despite composition, whether household or tradespeople. The community acts in the integrated system to avoid the extra cost to the system by identifying their advocate from the primary customer identification system. However, the system is capable of leading charity and donations inside the system, for example, transforming informal workers in the formal direction. The recycling waste management applies incentives and disincentives inside the system where indicates a condition that needs guidelines. However, this basic system adopts any further technology improvement in waste management infrastructure, hence avoid waste of resources and support waste management as a data tool for future analysis and research. Based on the smart identification, an affordable system developed by service blueprint shown in Fig. 7 as a model.

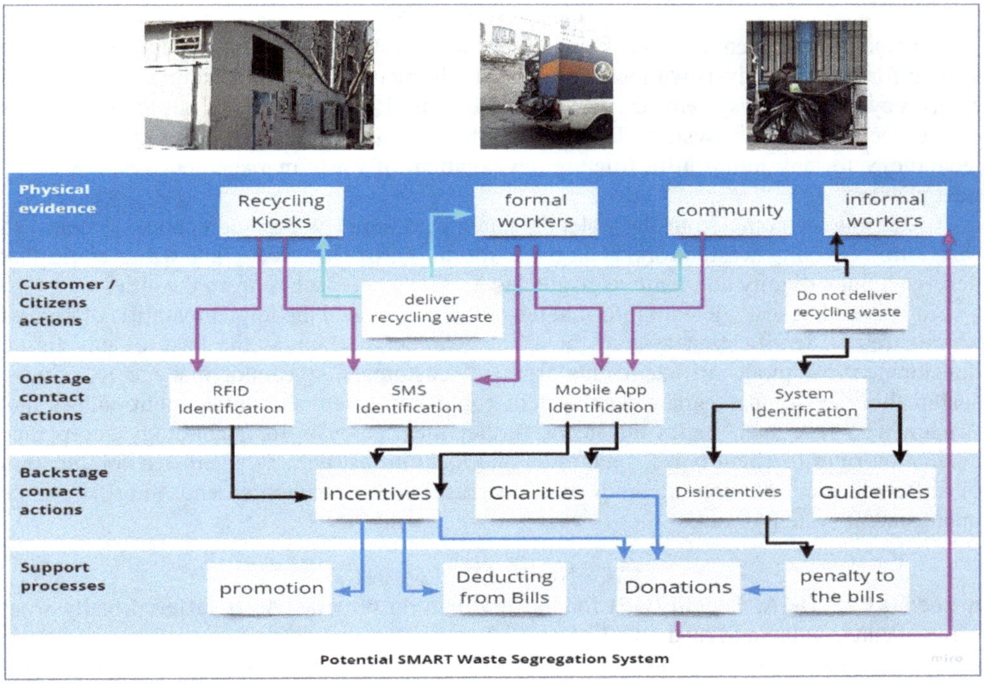

Figure 7: Potential smart recycling waste segregation.

3.2 Next research agenda

The current research is according to the analysis of customers' adaptation to the potential smart waste segregation system to identify the effects of the indicators on different aspects of Waste Management Dimensions environmental, economic and social where from Sanjeevi and Shahabudeen [5] point of view, in addition to two other political and administrative aspects, would be ideal for any city and simple enough for execution and performance evaluation and monitoring of an integrated SWM system. The proposed model demonstrates the administrative key role in developing a potential integrated system include system implementation, which followed by incentives, disincentive, and guideline application. Wherefrom the author's point of view, in the recycling waste system administrative aspect can merge to economic and social dimensions through incentives, disincentives, and guidelines. The political aspect evaluates inter-relation between different organizations that the issue could affect their boundaries like health and environmental organizations. Hence, any future research in this field can evaluate effective regulations and relation about the issue as agenda.

Though the technology acceptance model agenda evaluates potential system any further research on the performance of a plausible implemented integrated system recommended by the authors, therefore, the complementary study includes other aspects like financial analysis and technology acceptance model to iterate and improve.

4 CONCLUSION

The evaluation of potential smart system leads the waste management system to develop a system for further analysis with appropriate data. Implementation of a preliminary potential smart recycling waste system service serves the administration to figure out the procedures of the decision support system (DSS). Nevertheless, both systems need to address initial technology for any future infrastructure development of waste management systems in the next phase.

The resources come from the potential integrated smart waste segregation system can transform recycling waste workers into the formal part of the recycling waste circle that deserve human dignity and remove disgusting aesthetic perspective from the cities.

Any cities that struggle with informal recycling issues need the implementation of a smart waste stream identification system for all stakeholders, which is the base of any future development. Without sufficient data from all customers, it cannot manage to achieve sustainable development goals. The system could implement in similar metropolises like Ankara who face such issues but needs further local analysis for technology acceptance evaluation refer to client types, generation and local infrastructures whether a combination of different identification methods will increase their satisfaction and improve waste management system or not.

ACKNOWLEDGEMENT

In memory of Dr A. Sedghi (GP) for the information provided who suffered badly from Covid-19 in London and died on 27 May 2020.

REFERENCES

[1] United Nations, 68% of the world population projected to live in urban areas by 2050, says UN, 2018. www.un.org/development/desa/en/news/population/2018-revision-of-world-urbanization-prospects.html.

[2] United Nations, World population projected to reach 9.8 billion in 2050, and 11.2 billion in 2100, 2017. www.un.org/development/desa/en/news/population/world-population-prospects-2017.html.

[3] Olazable, M. & Pascual, U., Use of fuzzy cognitive maps to study urban resilience and transformation, *Environmental Innovation and Societal Transitions*, **18**, pp. 18–40, 2016.

[4] UN Environment Programme, Solid waste management, 2018. www.unenvironment.org/explore-topics/resource-efficiency/what-we-do/cities/solid-waste-management.

[5] Sanjeevi, V. & Shahabudeen, P., Development of performance indicators for municipal solid waste management (PIMS): A review. *Waste Management and Research*, **33**(12), pp. 1–14, 2015.

[6] Berger, C., Savard, G. & Wisere, A., EUGENE: An optimization model for integrated regional solid waste management planning. *International Journal of Environment and Pollution*, **12**(2/3), pp. 280–307, 1999.

[7] Gottingger, H.W., *Economic Models and Application of Solid Waste Management*, Gordon and Breach Science Publishers: New York, 1991.

[8] Huang, G.H., Baetz, B.W. & Patry, G.G., Grey dynamic programming for waste management programming under uncertainty. *Journal of Urban Planning and Development*, **120**(3), pp. 132–156, 1994.
[9] Abou Najm, M., El-Fadel, M., Ayoub, G., El-Taha, M. & Al-Awar, F., An optimization model for regional integrated solid waste management. 1: Model formulation. *Waste Management and Research*, **20**(1), pp. 37–45, 2002.
[10] Abou Najm, M. & EI-Fadel, M., Computer-based interface for an integrated solid waste management optimization model. *Environmental Modelling and Software*, **19**(12), pp. 1151–1164, 2004.
[11] Armijo, C., Puma, A. & Oieda, S., A set of indicators for waste management programs. *2nd International Conference on Environmental Engineering and Applications*, Singapore, 2011.
[12] Maryono, M. & Hasmantika, H., Preliminary study of smart urban waste recycling in Semarang, Central Java, Indonesia. *Proceedings of the International Conference of Smart City Innovation 2018*, 2019.
[13] Wilson, D.C. et al., *Global Waste Management Outlook*, United Nations Environmental Programme, 2015.
[14] Stickdorn, M. et al., *This is Service Design Doing: Applying Service Design Thinking in The Real World*, O'Reilly Media, USA, pp. 80–108, 2018.
[15] Citibrain, Value urban waste, 2019. www.citibrain.com/en/solutions/smart-waste/.
[16] Envac, 2019. www.envacgroup.com/.
[17] Sotkon Waste Systems, Intelligent systems, 2019. https://sotkon.com/en/intelligent-systems/.

DRY OLIVE POMACE GASIFICATION TO OBTAIN ELECTRICAL ENERGY IN A DOWNDRAFT GASIFIER

MARÍA DE LA TORRE MAROTO[1,2], JOSÉ ANTONIO LA CAL HERRERA[2,3] & MANUEL MOYA VILAR[1]
[1]Department of Chemical and Environmental Engineering (TEP 233),
Center for Advanced Studies in Energy and Environment (CEAEMA), University of Jaén, Spain
[2]Recursos Estratégicos de Biomasa, S.L., Spain
[3]Department of Business Organization, University of Jaén, Spain

ABSTRACT

Olive pomace is the main by-product generated by the olive oil mills, representing 80% of total fruit weight, which has high water content (60–70%). Reducing its moisture to 10–15%, the dry olive pomace (DOP) is obtained, which is a very interesting biofuel with a high calorific value (~4,100 kcal/kg). Annually, large quantities of this product are produced, nevertheless only part of it is used for self-consumption through electrical and thermal energy generation, by combustion. Olive oil mills are industries with high electrical energy consumption. The present study aims to give added value to this material, in order to supply electrical energy to the industries for self-consumption. The pelletized DOP (PDOP) has been subject to a gasification process using a downdraft fixed-bed gasifier in a pilot plant. The gas produced – syngas – has been cleaned through different filtering stages. Subsequently, it is introduced into a genset in order to generate electrical and thermal energy. The generated syngas has very promising characteristics, such as an upper calorific value (HHV) of 1,337.11 kcal/Nm^3 and a high content of nitrogen (N_2) and hydrogen (H_2), 47.29% and 19.54%, respectively. With this syngas, an average instantaneous power of 10 kW has been generated for 6 hours. The outcomes show the potential of using DOP as raw materials in the gasification process on an industrial scale. This allows to valorize the generated DOP in the olive oil mills and it contributes to reduce CO_2, it is favored the transition to the circular economy in the olive sector.

Keywords: dry olive pomace, biomass gasification, downdraft gasifier, syngas, power generation, self-consumption, circular economy.

1 INTRODUCTION

Andalusia is the main olive oil producer region worldwide. The olive grove area in Andalusia amounts to 1.5 million ha [1], which produces around 4.7 million tons of olives (Olea Europaea fruit) per year [2]. This olive is processed in the oil mills to make olive oil. The important activity developed by this sector results in the generation of a large number of by-products. Given the seasonality of the olive oil processing sector, the generation of this sub-product occurs in a very short time [1], between 4 and 5 months. Managing these agro-industrial by-products has become a problem for the mills, so their recovery is necessary.

Until the end of the 1980s, the olive oil production process was characterized by a three-phase centrifugation system, in which oil, water or "alpechín" and wet olive pomace (WOP) were produced. However, at the beginning of the 1990s, the three-phase process was changed to the two-phase process, in which, in addition to olive oil, a new phase is obtained, a combination of the other two that come out in the three-phase process. Therefore, the olive pomace, or "alperujo", produced contains all the water that previously came out as "alpechín" [3], [4]. Starting from a ton of olive, with 20% fat richness, 200 kg of olive oil and 800 kg of "alperujo" are produced, without adding water to the process. The "alperujo" has a high humidity (60–70%), which is why it is necessary to dry it to moisture between 10% and 15%, obtaining the dry olive pomace (DOP).

DOP has huge potential as a fuel; it is a homogeneous product with a high carbon content, high calorific value and low moisture. Annually, in Andalusia, an amount that ranges between

WIT Transactions on Ecology and the Environment, Vol 247, © 2020 WIT Press
www.witpress.com, ISSN 1743-3541 (on-line)
doi:10.2495/WM200131

1.2 and 1.45 million tons is generated [5]. A part of this production, approximately 20%, is invested in generating the thermal energy necessary for self-consumption by the industries themselves; another part, around 45%, is used as raw material to generate electrical energy in combustion plants; and the rest presents market difficulties at this time, being able to be valued through other alternatives [5]. The industries producing DOP, extractors of olive pomace oil, have a high demand for electrical energy and thermal energy, necessary for drying WOP.

Gasification is a thermochemical process that converts carbonaceous solid materials, such as biomass, into value-added gaseous products (syngas) [6]–[8]. In order for gasification to take place, a process initiating and maintaining agent is required. For this, steam and air are used, or this can be substituted by pure oxygen. In a typical gasification process the following stages take place: drying, pyrolysis, partial combustion and gasification [7], [8].

The different types of gasification processes are classified according to the type of reactor used, namely fluidized bed gasifiers and fixed bed gasifiers. After analyzing the different reactors, it was decided that the one that best suits the industries in the olive grove is the downdraft fixed-bed gasifier, since it is used for smaller units with powers equal to or less than 1 MW [9]. In these gasifiers, the gasifying agent is inserted above the reduction zone, flowing downwards and producing gasification in the lower part of the reactor and, therefore, the output of the gas produced through an ash bed at high temperatures [10]–[12]. These reactors produce less tar (0.015–3 g/Nm^3) than the rest, because the resulting gas is in optimal conditions for cracking to occur when passing through the hot ash bed [10], [11].

In this paper, a Downdraft fixed-bed gasifier is used. In this reactor the raw material is introduced through the upper part, while the gasifying agent enters through the lower part, at a certain height. The syngas produced flow down with the biomass exiting the bottom of the gasifier [10], as shown in Fig. 1.

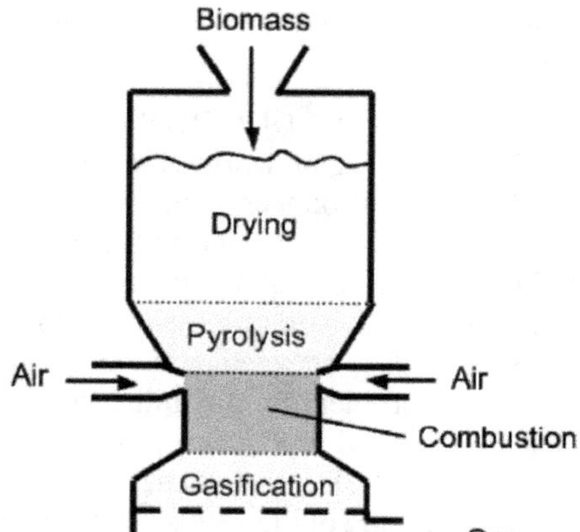

Figure 1: Schematic of a throated-type downdraft gasifier [9].

The aim of this work is the valorization of DOP, through its gasification in a downdraft gasifier. For this, previously, the olive pomace must be densified, due to its high content in fines (80%), through the use of a pelletizer, and then it is introduced into the gasifier. Everything will be done at the pilot plant level.

2 MATERIALS AND METHODS

2.1 Materials characterization

The wet olive pomace comes from an olive industry located in the province of Jaén, Spain. This olive pomace is dried in a rotary dryer, obtaining the raw material object of this study, DOP (Fig. 2(a)). The main physico-chemical characteristics of the product are shown in Table 1. DOP is a good fuel, due to its high calorific value (4,192 kcal/kg) and its high fixed carbon content (17.56%). However, about 80% of the sample has a particle size of less than 3 mm, therefore, this high content of fines can prevent fluidization during the gasification process.

To improve the biomass fluidization inside the gasifier, DOP is densified in a pelletizer, obtaining pelletized DOP (PDOP) (Fig. 2(b)) 10 mm in diameter and between 10 mm and 30 mm in length.

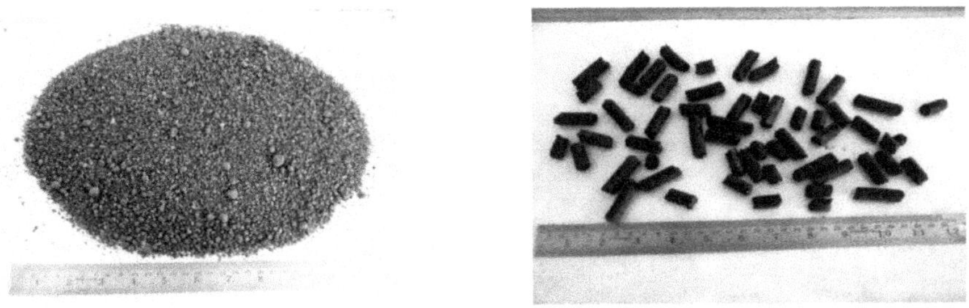

(a) (b)

Figure 2: (a) Sample of dry olive pomace; and (b) Pelletized dry olive pomace.

Table 1: Physico-chemical analyse of dry olive pomace and pelletized dry olive pomace.

Parameters	**Dry olive pomace**	**Pelletized dry olive pomace**
Moisture content, % wb	7.07	11.63
Ash, % db	9.38	7.36
Volatile, % db	73.06	71.77
Fixed carbon, % db	17.56	20.87
Ash fusion, in °C	–	at 1,200
Ash deformation, in °C	at 1,100	at 1,100
No Ash fusion, in °C	at 1,000	at 1,000
Bulk density, kg/m^3	580	680
True density, kg/m^3	–	1,393
Size, in mm	> 3mm: 20.48% < 3 mm: 79.52%	Diameter: 10 Length: 10–35
Ignition test	Burns easily	Burns easily
Flow ability test	Flows easily	Flows easily
Calorific value (HHV), kcal/kg db	4,192	4,429

www.witpress.com, ISSN 1743-3541 (on-line)

PDOP has better fuel characteristics than raw DOP and, in addition, allow an increase in the efficiency of the gasification process, while reducing gas treatment costs. The characterization of the PDOP can be seen in Table 1. When densifying, their calorific value and fixed carbon content are increased, reaching values of 4,429 kcal/kg and 20.87% respectively. Furthermore, its ash and volatile compounds content is reduced, going from 9.38% to 7.36% in the ash case and from 73.06% to 71.77% in the volatiles case.

For these reasons, it was decided to use the PDOP as raw material for the gasification process described below. For this, 5 tons of PDOP were prepared to send to India, where the pilot plant is located where the test will be carried out.

2.2 Experimental process

PDOP gasification plant consists of the downdraft fixed-bed gasifier, a gas cooling and cleaning stage, and an engine genset, all at the pilot plant level. An operation scheme is shown in Fig. 3.

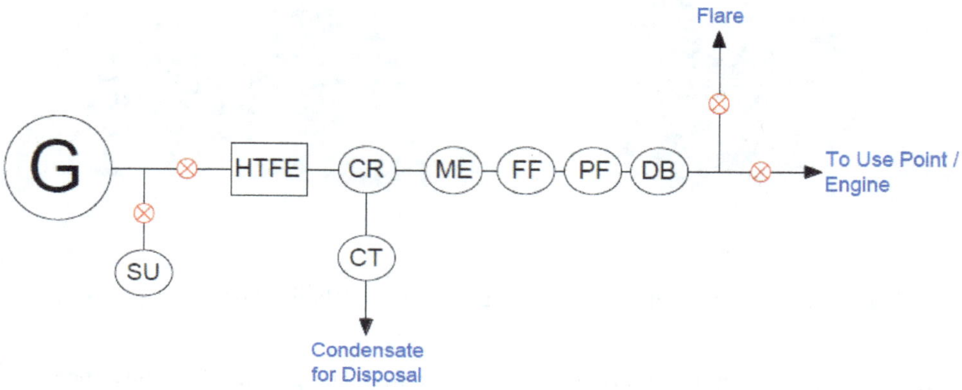

Figure 3: Gasifier schematic system.

The main stage of the process is the Downdraft fixed bed biomass gasifier "Ankur model FBG-100". The components of the gasifier, combustion cone, rotor, vertical grate, inner shell, nozzles, water flow channels and ash pipe are made of special stainless steel alloys, adapted to the conditions that must be met put up with. Some parts, such as hopper, structural support cones, etc., are made of mild steel with the necessary thicknesses and worked and painted appropriately. The entire blower construction is stainless steel.

After the gasifier, the ash removal system and the gas cleaning, dust, tar and inorganic impurities system is installed. This system is made up of high temperature filtering equipment (HTFE), condensate removal subsystem (CR), condensate neutralization subsystem (CT), mist eliminator (ME), fine filter (FF), pleated filter (PF) and, finally, dry blower (DB). The cyclone and scrubber system are also made of suitable stainless steel alloys. Filters are made of mild steel.

After the syngas have been cleaned, there is the engine genset, made up of two 100% gas engines, to produce electrical energy with a nominal power of 10 kWe and also thermal energy from cooling and exhaust gases.

3 RESULTS AND DISCUSSION

During the test implementation, the Downdraft gasifier was kept running for 9 hours. As a gasifying agent, air at atmospheric pressure was used. The biomass feed to the gasifier has been set at 75 kg/h. The temperature inside the gasifier ranged from 1,050–1,100°C. Conditions described above summarize in Table 2.

Table 2: Gasifier operating conditions.

Parameters	**FBG 100**
Total run hours	9
Gasifying agent	Air
Pressure	Atmospheric
Biomass consumption	75 kg/h
Temperature	1,050–1,100°C

3.1 Syngas production

The syngas characteristics at the gasifier outlet are measured, determining an average syngas flow of 156 Nm³/h at a variable temperature between 300°C and 500°C. The yield of the determined process has been 71%, in cold gas.

During the test, two gas samples are taken, at different times, to check the possible variation of its characteristics. The gas is collected filtered and clean. Table 3 shows the composition of both samples of the synthesis gas obtained from PDOP. It is observed that both gas samples have a very similar elemental composition with very small variations in the proportion (between 0% and 3%). The calorific value is also very similar in both samples, obtaining an average calorific value of 1,337.11 kcal/Nm^3.

Table 3: Produces syngas sampling, analysis and test results.

Gas parameters	**Unit**	**Sample 1**	**Sample 2**
Outlet gas temperature range	°C	300–500	
Average gas flow	Nm^3/h	156	
Cold gas efficiency of the system	%	71	
H_2	%	19.28	19.81
O_2	%	0.0	0.0
N_2	%	45.91	48.67
CH_4	%	2.74	2.42
CO	%	15.09	15.59
CO_2	%	16.98	13.51
Calorific value (HHV)	kcal/Nm^3	1,336.11	1,338.12
Average HHV	kcal/Nm^3	1,337.11	

3.2 Electricity production

The produced and clean syngas is introduced into a genset consisting of two 100% syngas engines, to produce electrical energy with a maximum gross power of 10 kWe and also thermal energy from cooling (hot water at 90°C) and from the exhaust gases at approximately 400°C, also usable. During the operation of both motors, every 30 min or one hour, the

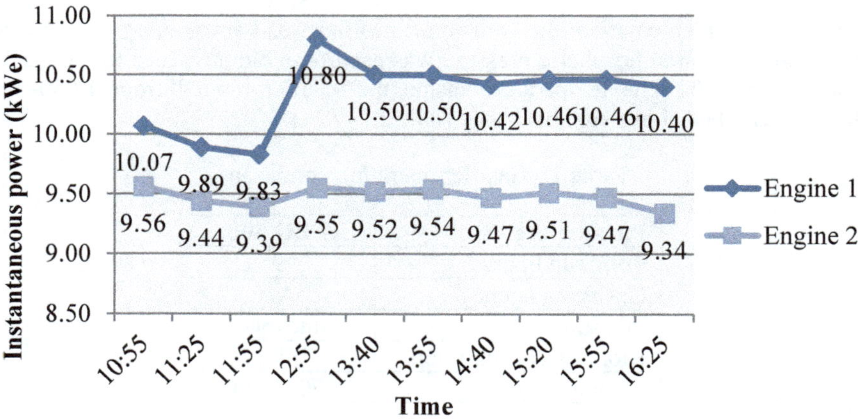

Figure 4: Instantaneous power evolution during the test.

Figure 5: Discharge bio-char photography.

instantaneous power produced is measured, whose measurements can be seen in the graph in Fig. 4. Motor 1 obtains higher instantaneous power values with an average of 10.33 kWe, while motor two gets an average of 9.47 kWe.

3.3 Bio-char production

A solid product called bio-char (Fig. 5) is also obtained from the gasification process, with an average production of 10.77 kg/h. This represents 14.36% of PDOP consumption.

Table 4 shows the bio-char physico-chemical characteristic. As can be seen, it has a high moisture content (38.38%), which is due to the sum of the residue's own moisture, plus that added during discharge in order to decrease its temperature at the gasifier outlet. The characteristics of the bio-char are very good for use as a fuel, since it has a high fixed carbon content (66.58%) and a high calorific value (5,841 kcal/kg), on a dry basis. Furthermore, its volatile content (6.93%) is reduced compared to the content present in the raw material (71.77%). However, it has a high ash content (26.49%), which can pose a problem for the cleaning and maintenance of combustion equipment.

Table 4: Physico-chemical analyses of discharge bio-car.

Parameters	**Bio-char**
Average dry discharge	10.77 kg/h
Moisture content, % wb	38.38
Ash, % db	26.49
Volatile, % db	6.93
Fixed carbon, % db	66.58
Bulk density, kg/m^3	675
Size, in mm	0–20
Ignition test	Burns easily
Flow ability test	Flows easily
Calorific value, kcal/kg db	5,841

4 CONCLUSION

From the results obtained, it can be deduced that the gasification of the PDOP is a good alternative for the recovery of this kind of waste, since a high gas yield with a high calorific value is obtained, and a carbonaceous residue that can have a very good market. Furthermore, with the PDOP gasification the following advantages are obtained: reduction of energy costs (OPEX) and reduction of CO_2 emissions in the olive oil sector; favoring the transition towards a circular economy in this sector. Sector well established in the Mediterranean countries, highlighting above all Spain.

ACKNOWLEDGEMENTS

The authors thank the University of Jaén and Recursos Estratégicos de biomasa S.L. for financing this work through the "Specific collaboration agreement between Recursos Estratégicos de Biomasa S.L. and the University of Jaén for the development of a doctoral project applied in the co-financing modality" dated 11 Sep. 2018.

REFERENCES

[1] Agencia de Gestión Agraria y Pesquera de Andalucía, Junta de Andalucía (AGAPA), Evaluación de la producción y usos de los subproductos de las agroindustrias del olivar en Andalucía. Unión Europea: Fondo europeo Agrícola de Desarrollo Rural, 2015. www.juntadeandalucia.es/agriculturaypesca/observatorio/servlet/FrontController?action=DownloadS&table=11031&element=1585173&field=DOCUMENTO. Accessed on: 5 Jun. 2020.

[2] Consejería de Agricultura y Pesca (CAP), Junta de Andalucía, Potencial energético de los subproductos de la industria olivarera en Andalucía. 2010. www.juntadeandalucia.es/export/drupaljda/Potencial%20energ%C3%A9tico.pdf. Accessed on: 5 Jun. 2020.

[3] Roig, A., Cayuela, M.L. & Sánchez-Monedero, M.A., An overview on olive mill wastes and their valorisation methods. *Waste Management,* **26**, pp. 960–969. DOI: 10.1016/j.wasman.2005.07.024.

[4] Alburquerque, J.A., Gonzálvez, J., García, D. & Cegarra, J., Agrochemical characterisation of "Alperujo", a solid by-product of the two phase centrifugation method for olive oil extraction. *Bioresource Technology,* **91**, pp. 195–200, 2004. DOI: 10.1016/S0960-8524(03)00177-9.

[5] Agencia Andaluza de la Energía (AAE), Consejería de Hacienda, Industria y Energía, Junta de Andalucía, La bioenergía en Andalucía, 2020. www.agenciaandaluzadelaenergia.es/es/biblioteca/la-bioenergia-en-andalucia. Accessed on: 5 Jun. 2020.

[6] Basu, P. (ed.), Introduction. *Biomass Gasification and Pyrolysis: Practical Design and Theory*, Academic Press, pp. 1–25, 2010. DOI: 10.1016/B978-0-12-374988-8.00001-5.

[7] Basu, P. (ed.), Gasification theory and modeling of gasifiers. *Biomass Gasification Design Handbook*, Academic Press: Boston, pp. 117–165, 2010. DOI: 10.1016/B978-0-12-374988-8.00005-2.

[8] Ciuta, S., Tsiamis, D. & Castaldi, M.J. (eds), Fundamentals of gasification and pyrolysis. *Gasification of Waste Materials,* Academic Press, pp. 13–36, 2018. DOI: 10.1016/B978-0-12-812716-2.00002-9.

[9] Situmorang, Y.A., Zhao, Z., Yoshida, A., Abudula, A. & Guan, G., Small-scale biomass gasification systems for power generation (<200 kW class): A review. *Renewable and Sustainable Energy Reviews*, **117**, 109486, 2020. DOI: 10.1016/j.rser.2019.109486.

[10] Basu, P. (ed.), Design of biomass gasifiers. *Biomass Gasification, Pyrolysis and Torrefaction: Practical Design and Theory*, Academic Press, pp. 167–228, 2010. DOI: 10.1016/B978-0-12-374988-8.00006-4.

[11] Basu, P. (ed.), Design of biomass gasifiers. *Biomass Gasification, Pyrolysis and Torrefaction: Practical Design and Theory*, Academic Press, pp. 263–329, 2018. DOI: 10.1016/B978-0-12-812992-0.00008-X.

[12] Guangul, F.M., Sulaiman, S.A. & Ramli, A., Gasifier selection, design and gasification of oil palm fronds with preheated and unheated gasifying air. *Bioresource Technology* **126**, pp. 224–232, 2012. DOI: 10.1016/j.biortech.2012.09.018.

IN-SITU BACKFILLING EXPERIMENT OF THE SMALL SCALE DRIFT BY SPRAY METHOD IN MIZUNAMI UNDERGROUND RESEARCH LABORATORY, JAPAN

HIROYA MATSUI[1], RYOJI YAHAGI[2], HIKARU ISHIZUKA[2] & TOMOHITO TOGURI[2]
[1]Sector of Nuclear Fuel, Decommissioning and Waste Management Technology Development, JAEA, Japan
[2]Shimizu Corp., Japan

ABSTRACT

In-situ backfilling experiment using spray method in the small scale drift (approximately 4 m by 3 m scale) was conducted at a 500 m depth in the Mizunami Underground Research Laboratory (MIU), established by JAEA (Japan Atomic Energy Agency). The aim of this experiment is to demonstrate backfilling by spray method deep underground and establish the quality control methodology as the one of the techniques applied for geological disposal in Japan. The backfill material consists of a sand and bentonite mixture. Specification for the backfill material was designed to satisfy the target permeability of a generic host rock (10^{-8} m/s) assumed by NUMO. In this case, effective clay density should be 0.4 Mg/m^3 or more. Quality control of the material before backfilling was performed by setting the initial water contents (14±2%) based on the results of the Laboratory testing and preliminary spray testing on ground surface. While the quality control of the backfilled material during the experiment was performed by density measurement with the standard method and a 3D-scanning method. Densities of the backfilled material measured at any points satisfied the specification (more than 0.4 Mg/m^3) and the results suggested the establishment of the practical quality control methodology of the backfilling by spray method under actual deep geological environment. Furthermore, the measurement of the moisture transfer characteristics and the swelling of bentonite using several instruments, as well as the amount of the outflowing of bentonite from the backfilled material, was carried out.

Keywords: backfilling, spray method, in-situ experiment, deep underground, crystalline rock.

1 INTRODUCTION

One of the features of the geological disposal policy in Japan is the establishment of multiple URLs. The URLs must be distinct from actual disposal facilities which will be constructed by the Nuclear Waste Management Organization of Japan (NUMO). The URLs of the JAEA projects are distinct from on-site (site-specific) URLs classified as purpose-built generic URLs as described in the OECD/NEA report [1], to be constructed at potential waste disposal sites. JAEA's URL projects aim to improve the reliability of geological disposal technologies and to develop advanced safety assessment methodologies. In order to characterize the general range of geological environments in Japan, two generic URLs were constructed. One is the Mizunami Underground Research Laboratory (MIU) focused on crystalline rock (Fig. 1), and the other is the Horonobe Underground Research Laboratory focused on sedimentary rock. The MIU project has been performed in three overlapping phases: "Phase I: Surface-based investigation," "Phase II: Construction" and "Phase III: Operation." The construction of research galleries in the MIU has completed to –500 m depth. Since 2010, research activities in Phase III have started in the underground. Current R&D activities include a gallery closure test at the GL-500 m level to study recovery processes in the geological environment around a gallery after it has been backfilled, the development of long-term monitoring technology, a post-excavation grouting experiment to demonstrate the feasibility of impermeable grouting and a mass transport experiment in MIU [2].

WIT Transactions on Ecology and the Environment, Vol 247,
www.witpress.com, ISSN 1743-3541 (on-line)
doi:10.2495/WM200141

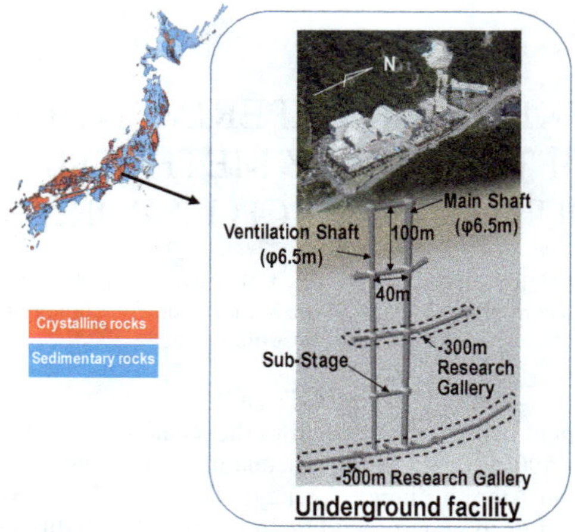

Figure 1: Location of MIU and schematic layout of the gallery [2].

The in-situ backfilling experiment was carried out as a part of development of technology for backfilling the underground structure in MIU excavated in crystalline rock as one of the current R&D items.

2 SITE LOCATION AND GEOLOGICAL ENVIRONMENT

The MIU Construction Site is located at an altitude of 200 m above sea-level in a hilly area adjacent to the urban area of Mizunami City, Gifu Prefecture, central Japan. Around the site, Tertiary sedimentary rocks widely and unconformably overlay the cretaceous Toki Granite (60–70 Ma) [3]. The Toki Granite, a zoned pluton, has three rock facies grading from muscovite-biotite granite at the margin through hornblend-biotite granite to biotite granite [3]. The R&D work of the MIU project conducted mainly in the Toki Granite basement.

Due to the geological disposal of HLW will be under 300 m depth according to Japanese law, experiment site shown in Fig. 2 was located in research gallery excavated 500 m depth as a demonstration. The site was a horseshoe-shaped cross-section drift with a length of about 10 m, a height and a width of about 3 m. Inflow rate in the drift was less than 0.1 L/min and it is relatively small in MIU drift in granite.

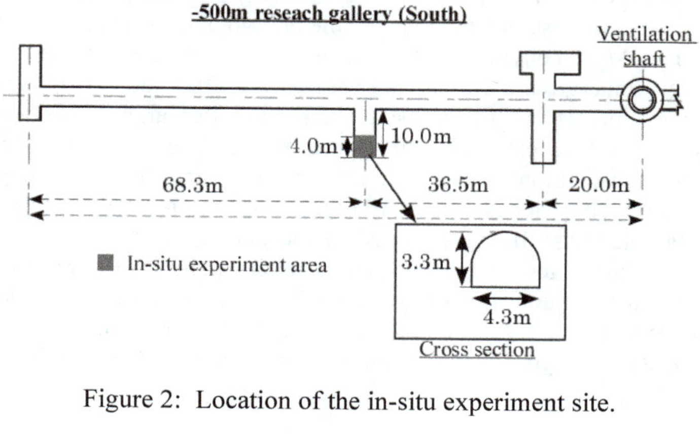

Figure 2: Location of the in-situ experiment site.

3 PLANNING

The Backfilling experiments were carried out both within and outside Japan. Geological disposal in Japan is still R&D phase for geological disposal and engineered barrier system related to the backfilling is not fixed yet. The experiment for horizontal disposal technology has been conducting at Horonobe URL. Therefore, backfilling is assumed the type of KBS-V in this experiment and focused on backfilling of disposal tunnel.

We referred previous studies in Japan and overseas that were subject to literature surveys for planning. Literature surveys were carried out in view of the site-generic perspective in Japan for geological disposal and the in-situ experiment plan was focused on backfilling a drift, in the case of vertical deposition of the MIU project. As a result, advection suppression, chemical stability between materials, buffer position retention, canister lift prevention and tunnel stability of disposal tunnels were extracted as the required performance for backfilling. In addition, the result confirmed that the permeability and swelling pressure of the backfill material were used as performance indicators. These results are the same as required performance for backfilling that NUMO has studied.

From the viewpoint of a site-generic test plan, it was determined that the values of the permeability of backfill was controlled to be less than the one of a host rock presented in NUMO [5]. Specifically, measured averaged hydraulic conductivity of Toki granite around in-situ experiment site in Fig. 3 was 10^{-8} m/s order and the permeability of backfill was less than 10^{-8} m/s order as a target.

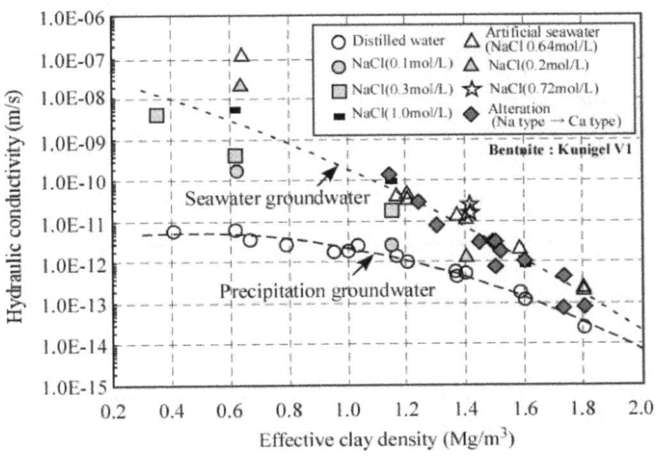

Figure 3: Relationship between hydraulic conductivity and effective clay density [5].

Fig. 3 shows the relationship between the density of the clay material (bentonite) to be applied for geological disposal (effective clay density) and permeability. The result suggested that the effective clay density over 0.4 Mg/m^3 satisfies the permeability of 10^{-8} m/s or less for both groundwater and saline water. Therefore, effective clay density of 0.4 Mg/m^3 or more was adopted as a control parameter for the material specifications in this experiment. Effective clay density (ρ_e) is calculated following equation:

$$\rho_e = \frac{\rho_d(100 - R_s)}{(100 - \frac{\rho_d R_s}{\rho_s})}$$

where ρ_d: dry density (Mg/m^3); R_s: mixture rate of dry sand; ρ_s: soil particle density of sand. (Average value of dry density of sand and crushed rock in the case of crushed rock is used.)

On the other hand, it was judged that this test will be able to conduct for only two years due to political situation of MIU project. Therefore, tunnel stability of disposal tunnels, the one of the required performance was not adopted in this study because the presence or absence of backfilling material does not affect the mechanical stability of the in-situ experiment site. In addition, it was decided that a retaining wall that allows drainage was constructed instead of a plug due to the very short period allowed for this experiment.

Based on above condition, the preliminary Lab. testing using different compaction energy of mixed soil composed of bentonite (Kunigel V1), sand, and crushed rock was carried out as shown in Table 1 and Table 2. The results show that both materials were satisfied the above described requirement (effective clay density is 0.4 Mg/m^3 or more) under different compaction energy.

Table 1: Specification of the tested backfilling material.

	Composition	Mixture(wt%)	ρ_s(Mg/m^3)
Material A	Kunigel V1	15	2.767
	Sand	35	2.681
	Crushed rock	50	2.686
Material B	Kunigel V1	15	2.759
	Sand	85	2.674

Table 2: Results of laboratory compaction testing.

	Material A			Material B		
Compaction energy	1Ec	0.6Ec	0.4Ec	1Ec	0.6Ec	0.4Ec
$\rho_{d\,max}$(Mg/m^3)	1.933	1.855	1.799	1.747	1.681	1.648
w_{opt} (%)	11.6	12.5	14.8	14.2	14.5	15.9
ρ_e (Mg/m^3)	0.727	0.656	0.609	0.589	0.542	0.519

There are several construction methods of backfill material, which are rolling and compaction, block stacking and spraying and they were demonstrated in Japan and overseas. In this experiment, a full-section spraying method was applied, which is the unprecedented method within and outside Japan. However, there was large limitation of the materials and machines used for the spray construction because they were required to pass through the opening (1.9 m) of the ventilation shaft scaffold of MIU. Thus, examine the quality of backfill materials was planned for the development of quality control methodology, the full-section spraying method. Consequently, the backfilling in this experiment was planned to divide into 8 steps, and the measurement work and the spraying construction are repeated for each step (Fig. 4). Table 3 shows the measurements for quality control. These items were planned based on the results of literature survey as well.

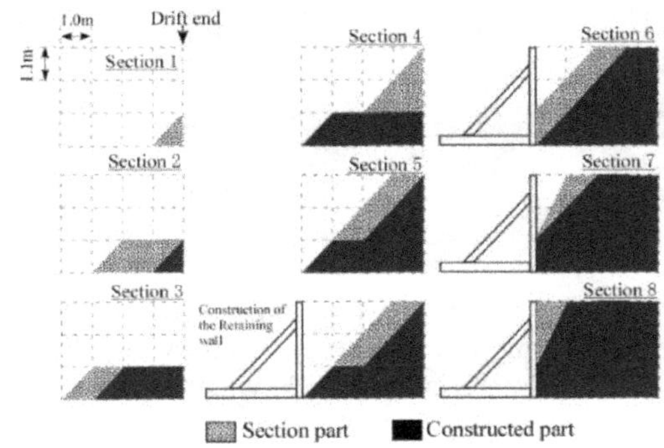

Figure 4: Steps of backfilling in this experiment.

Table 3: Measurement and monitoring items.

Item	Method	Number	Objective
Density	Lab. Test	50	Confirmation of the quality of sprayed buffer material
	Evaluation based on the 3-D scanning data	After spraying of each section	
Total Pressure	Pressure meter	10	Mesurement of the swelling pressure
Water Pressure	Water pressure meter	12	Mesurement of the water pressure
Water content	Soil moisture meter	10	Variation of water content in buffer material
Total outflow volume	Water lebvel indicator	1	Estimation of the outflow of bentnaite from buffer material
Concentration of bentnaite in outflow	Absorbance meter	1	

4 SELECTION OF MATERIALS AND MACHINES FOR THE IN-SITU BACKFILLING EXPERIMENT

According results of planning, the backfill material A and backfill material B were used to investigate the effects on workability and quality due to the difference in materials used for in-situ spray on ground surface (Fig. 5).

Firstly, the test was conducted by spraying of backfill material A (the water content ratio is set to 13.7% on average of 0.4Ec and 0.6Ec optimum water content ratio) using a spraying machine (Ariba 285). The result shows that the effective clay density after spraying was 0.571 to 0.611 Mg/m^3, while a relatively high density was obtained. On the other hand, it required 63 minutes (net spraying time of 13 minutes) for spraying one bag (428 kg). The main reason was that the addition of crushed rock. It made the backfill material agglomerated by compaction during transportation and it was difficult to put into the spraying system. Therefore, material A was not adopted for this experiment due to practical reason.

(a) (b) (c)

Figure 5: Preliminary test on ground surface. (a) Layout of preliminary test on ground surface; (b) Spraying the buffer material with 14% water content; and (c) Spraying the buffer material with 18% water content.

The following test was conducted with material B. In this test, equipment and machine were selected taking into account workability and quality based on the result of the first test shown in Table 4. Especially, spraying machine (AGC PRIBRICO with a small spraying amount per injection) was used to prevent clogging in the material hose. In this test, the requirement was set to be up to the dry density after sprayed material up to that of 0.4Ec in Lab. conservatively. Fig. 6 shows the dry density of material B after sprayed with 14% and 16% water content was expected up to that in the case of 0.4Ec. Regarding workability (Table 4), it was confirmed that a net spray amount of about 1 ton per hour could be secured in the cases of w=14% and 18%. This result shows that improvement from the previous test. The minimum loss rate was obtained at W=14%. On the other hand, stability of sprayed material and density were extremely low in the case of W=18% (shown in Fig. 6).

Table 4: Obtained speed of spray method on ground surface.

	W=14%	W=16%	W=18%
Total time(min)	62	79	53
Time of spraying(min)	29	31	20
Weight of sprayed material(kg)	951	858.5	820.5
Speed of backfilling by spraying (kg/h)	1967.6	1661.6	2461.5
Loss rate*	1.287	1.445	1.364

* Loss rate: Total weight of used material / Weight of sprayed material

Based on above series of selection processes and test results, materials and machines were selected for the in-situ experiment. Table 5 shows the material specifications of backfill materials. The moisture content of backfill material was set to 14 ± 2% (considering errors due to mixing), which resulted in high construction density and good workability (construction speed, loss rate).

5 OVERVIEW OF THE IN-SITU BACKFILLING EXPERIMENT

The demonstration test was conducted in the equipment horizontal shaft (GL-500 m) of Mizunami Underground Research Laboratory. Example of the state of in-situ spraying is shown in Fig. 7.

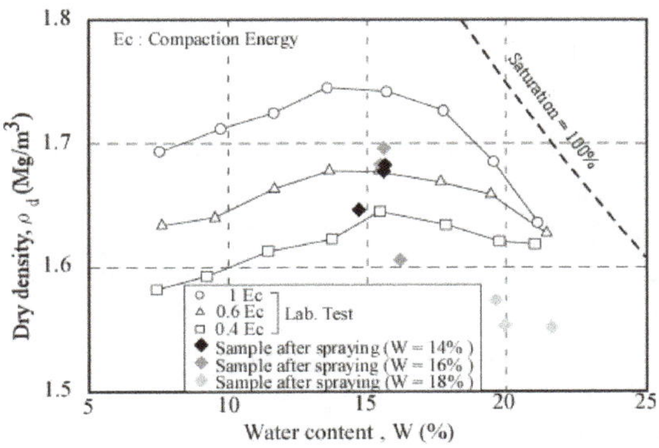

Figure 6: Results of the compaction test with different water content (Lab. test and preliminary test on the ground surface for material B).

Table 5: Final mixture of backfilling material for in-situ experiment.

	Mixture (%)	Water content (%)
Bentnaite (Kunigel-V1)	15	14±2
Sand	85	

Figure 7: Spraying of back fill material.

In this test, construction speed and spraying speed were applied for workability evaluation, and effective clay density (ρ_e) was applied as an index for quality evaluation. Fig. 8 show the water contents of the material at site preparation and after spraying. Most of the material water content just before spraying were satisfied with its specification (14±2%) and well controlled. And the water content after spraying was also in the same range. The

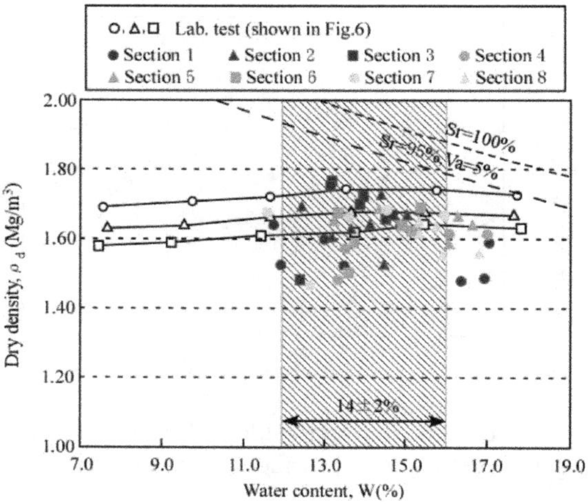

Figure 8: Measured water content of buffer material at site preparation.

Table 6: Results of the evaluated backfilling speed in each section and estimated effective clay density of backfilling material with different methods.

	Weight of Sprayed buffer material(kg)	Speed of backfilling by spraying (kg/h) (Ave.)	Effective bentnite density (Mg/m3)	
			Lab.test (Ave.)	Evaluation based on the 3-D scanning data
Preriminary testing at ground surface	951.0	1967.6	0.534	-
Section 1	6173.0	1119.0	0.447	0.326
Section 2	13689.7	1342.1	0.527	0.450
Section 3	8308.1	1724.9	0.566	0.514
Section 4	22394.4	1514.8	0.477	0.420
Section 5	13542.6	1599.5	0.521	0.408
Section 6	16319.5	1962.3	0.509	0.343
Section 7	9115.8	1905.7	0.513	0.587
Section 8	5015.3	1662.5	0.492	0.631

measurement for material density, water content etc. of sampled backfill material were performed at on-site immediately.

Table 6 shows the summary of the results of averaged construction speed by in-situ spraying and density at each section. Measured density except existing gutter part showed 0.41–0.60 Mg/m^3 of effective clay density and it is confirmed to satisfy the requirement of ρe (0.4 Mg/m^3 or more). The averaged backfilling speed by spray method in in-situ obtained

1119.0–1962.3 kg/h. The speeds of the first 2 sections were relatively low due to inexperience of workers. However, the speed after section 2 is close to it of second test on ground surface and the total system for in-situ spraying was concluded with productive work in this experiment.

Fig. 9 shows the planed and actual sampling points of splayed material. Coordination of actual sampling points were obtained from 3D scanning data. The planned sampling positions were determined with consideration of the reliable results of 3-D contour map. The difference between detail planed and actual positions was mainly due to practical reason. Fig. 10 shows the 3-D distribution of effective clay density of backfill material using measured actual sampling position as an example. The results suggested that the effective clay density close

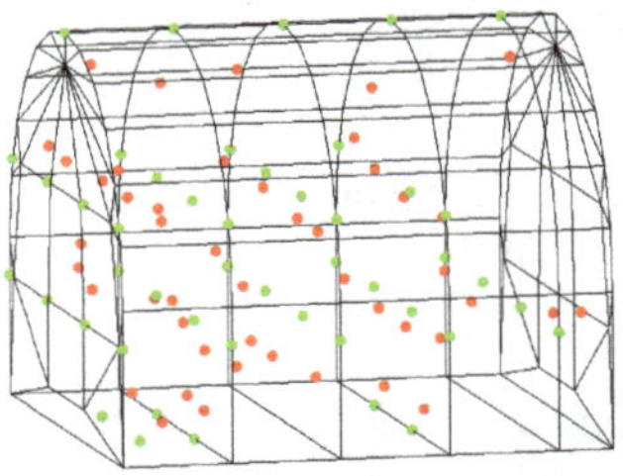

Figure 9: Sampling points (planned: red; actual: green).

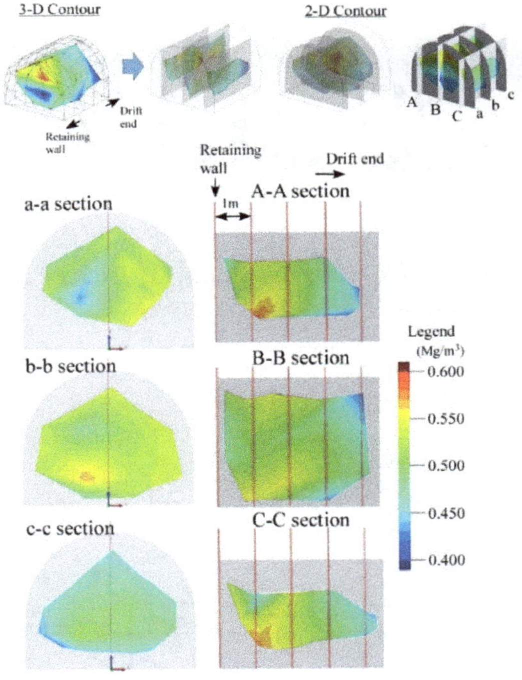

Figure 10: An example of distribution of the measured properties of backfill material based on 3-D contour map (effective clay density).

to drift end was relatively low, Section 1 and Section 6 in this case. In Section 1, the splaying pressure might be low because of inexperience of workers and energy loss by upward splaying affect for the density in Section 6.

Fig. 11 shows an example of the 3D scanning results. 3D-scanner was Focus3D S120(FARO Co.ltd.) and averaged time for 3D scanning in this experiment was about 20 min/1 scanning. Fig. 12 shows the comparing of the effective clay density between measured and calculated based on 3D scanning data and actual weight by spraying. Error bar shows the variation of the measured effective clay density in each section. Variation range of measured effective clay density was max. 0.1 Mg/m^3. The effective clay bulk density of Sections 1–6 wa smaller than that of measured. It might be caused by the overestimate of the volume due to insufficient extract of rebound at those sprayed parts. While effective clay bulk density of the Sections 7 and 8 was higher, which was caused by the underestimation of the sprayed volume due to some obstacles (retraining wall etc.). The estimation of the bulk density based on 3D scanning has several advantages (no damage at backfilled part, minimize the impact on the backfilling schedule because of its short-term measurement and safety of backfilling work) and it may be useful if the calibration curve like indicated in Fig. 12 will be made in the future geological disposal.

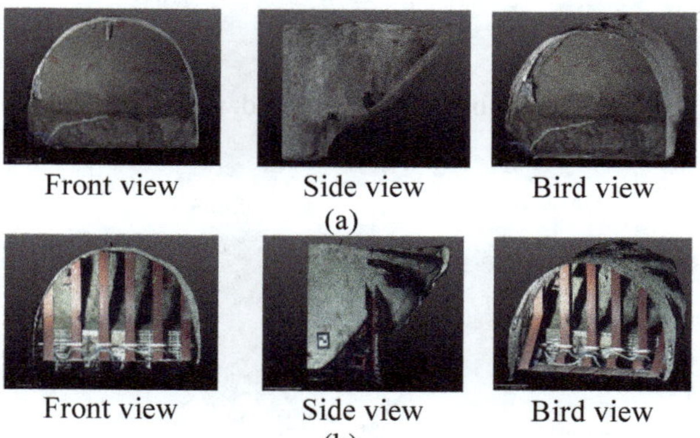

Figure 11: An example of 3D scanning results. (a) Section 5; and (b) Section 6.

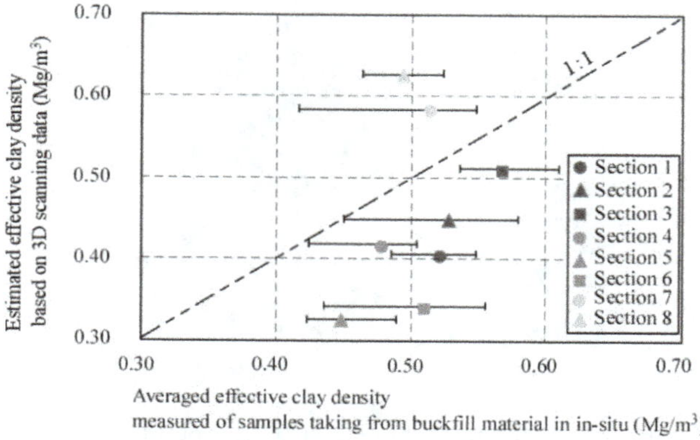

Figure 12: Relation between estimated effective clay densities.

Fig. 13 shows the used equipment, arrangement of monitoring during and after backfilling. Fig. 14 shows the schematic layout for the in-situ measurement of erosion of bentonite from backfill material. Monitoring results are summarized below.

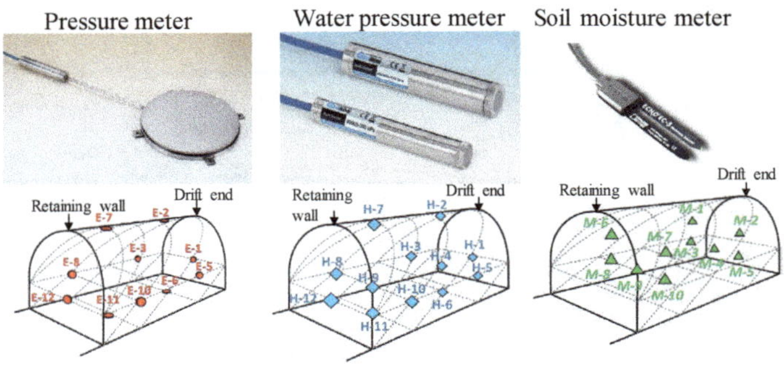

Figure 13: Monitoring equipment and its arrangement.

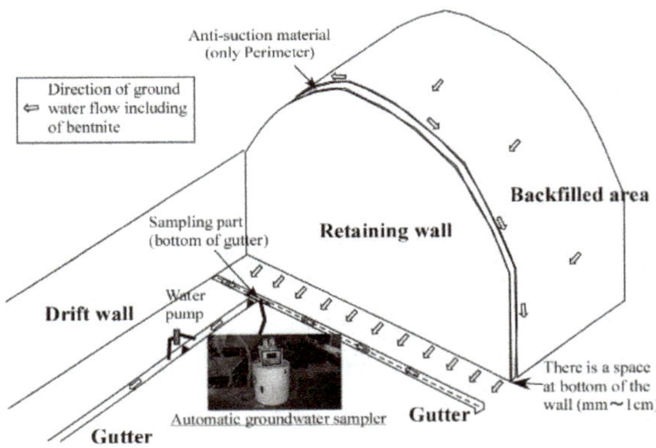

Figure 14: Monitoring of the outflow of ground water including of a bentonite.

The soil moisture meter (M-1, M-2, M-6, M-7) installed on the top end shows a value of 80% or more of water content immediately after installation, and those parts maintain stable condition. Moisture in other backfill materials also gradually rose until around May 2019, indicating a moisture content of more than 60%. The soil moisture meter at the M4 point showed a value of 100% immediately after installation; however, the value suddenly dropped in the middle of February 2019 then showed unstable behavior. Possible cause of the behavior is failure of measuring instrument (Fig. 15(a)).

Most of water pressure gauges resulted in a slight increase in negative pressure over time. On the other hand, qualitatively, H-1 and H-5 that are closest to the face turned to positive pressure, which is consistent with the infiltration of groundwater from the boundary of backfilled area and drift wall (Fig. 15(b)).

For total earth pressure, the E-11 point on the bottom and the E-1 point on the face tend to rise slowly until May 2019 while fluctuating. On the other hand, the point at E-10 tends to decrease gradually. Other than that, it rose immediately after installation and then levelled off (Fig. 15(c)). E12 installed on the retaining wall has been almost staying at the same value and there is no appearance of swelling pressure behavior on the retaining wall. It was not able to extract the swelling pressure of bentonite in total earth pressure quantitatively due to its quite small value (averaged swelling pressure predicted based on the previous study result was 0.013 MPa [6]).

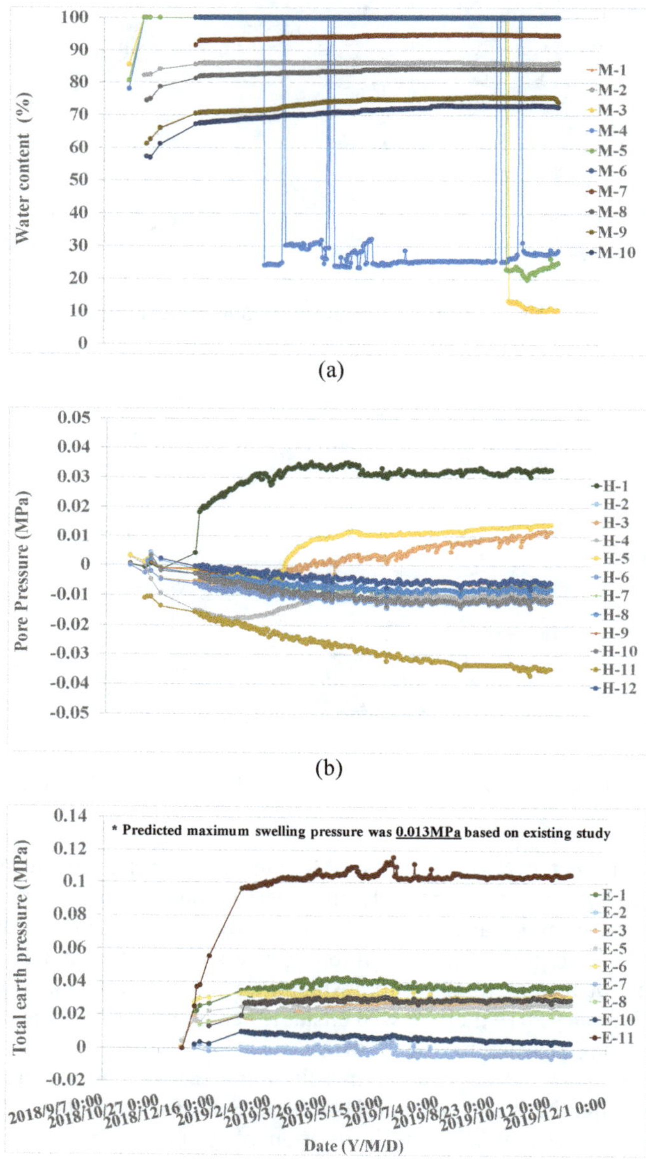

Figure 15: Monitoring results. (a) Soil moisture; (b) Water pressure; and (c) Total earth pressure.

From the variation in the total earth pressure, the state changes during the backfilling can be roughly divided into two stages (Stage 1: Start – middle of May, Stage II: Middle of May – Monitoring end). The state and its possible reason in each stage as follows:

Stage I: All earth pressures including the two earth pressure at the top increase gradually and suddenly drop at some point. During this period, soil moisture also showed a relatively increasing trend, and the water pressure also increased to positive pressure in some cases. It is supposed that the degree of saturation increases and the swelling behavior occurs due to the infiltration of groundwater into the backfill material, however, when the pressure behavior on the drift wall reaches the certain level, the pressure drops rapidly due to some deformation in backfilling part.

Stage II: Moisture rise almost constantly and the total earth pressure moves up and down on the spike. Water pressure shows a tendency to being stable. The infiltration of groundwater in the test area has reached the steady state (not full saturation). Flow paths through, which may be made by backfill material. Changes in the spike of the earth pressure seem to be linked to the behavior of groundwater infiltrated in the backfill material through the paths.

It is estimated that above different conditions are observed due to the fact that outflow from backfilling part is allowed for the in-situ measurement of erosion of bentonite and the local heterogeneity of the physical properties in the backfill area.

Fig. 16 shows the comparing of the bentonite erosion between the results of an existing study [7] and this experiment. The results are consistent with existing study for Kunigel V1 bentonite. In addition, amount of erosion volume with total outflow is also similar with SKB's results using MX-80. Although the bentonite used and experiment condition were different, it is interested and the results of our experiment will provide useful information for predicting the bentonite erosion when the future geological disposal is designed.

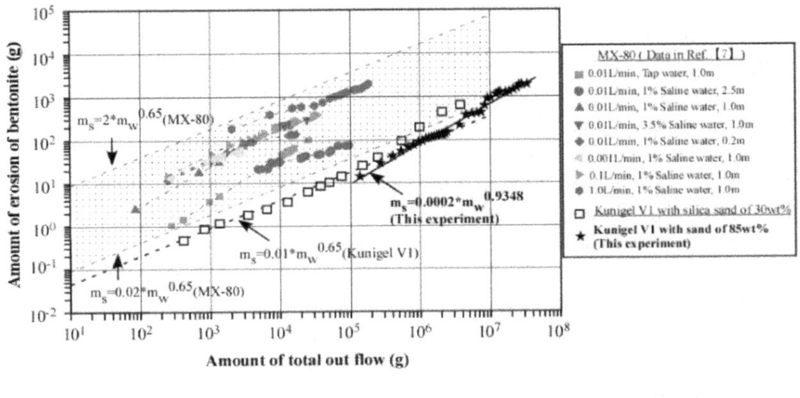

Figure 16: Comparing between existing results and the result of in-situ measurement for erosion of clay (bentonite) in the backfilled material of this study.

6 CONCLUDING REMARKS

In this in-situ experiment, the applicability of spray method for backfilling was evaluated from both the "workability" and "quality" aspects. In the evaluation of workability, parameters directly related to the construction work such as "spraying speed" were calculated, and “effective clay density” was used as an evaluation index of “quality”. The target value of the effective clay density is set to 0.4 Mg/m^3, which correspond with the averaged permeability of host rock around test site (10^{-8} m/s order). It is difficult to compare the construction results in other places in Japan and overseas and the applicable construction methods (rolling and compaction, block construction, etc.), because of the various geological environment and the environmental conditions of a place where it can actually be used. However, the full-section spraying method applied in this experiment can be evaluated as one of the practical methods for backfilling in the future geological disposal project for the following reason:

- The spraying speed, which are the parameters of “workability,” are comparable to the results of preliminary tests on the ground surface where there are almost no restrictions except for the Section 8, that was final part.
- Regarding the “quality” of the backfill material required, all sections can achieve a value up to 0.4 Mg/m^3 or more which required low permeability (at least 10^{-8} m/s order of host rock).

On the other hand, from the data of construction in this in-situ experiment, the operator’s experience, spraying conditions (nozzle position and orientation) and construction environment (ventilation environment and sufficient electric energy) can affect the workability and quality. In particular, the available utilities in underground such as electricity, ventilation system etc. may become a practical problem even in an actual geological disposal site and it is necessary to consider the design of the future geological disposal facility. The decreasing of spraying pressure was observed when spraying with the nozzle facing upward. It may be resulted in decreasing of density of the backfill material. In this case, the control of the nozzle direction keeping either horizontal or downward as possible and the improvement of the spray nozzle and sufficient discharge volume at least for the height of the backfill area (3 m in this test) were performed. As a result, the influence was minimized.

In addition, after spraying, a slight gap of a level was observed near the top, which caused the self-weight settlement with time of the backfill material mainly and backfilled material close to the drift wall was loosed by only a small amount of ground water from the drift surface. It was presumed that the backfill material was made near the optimal moisture content for compaction. This also causes the degraded quality of the backfill material. It seems that one of the countermeasures is to use a backfill material with a high bentonite content at spraying close to drift wall and it is desirable to incorporate the use of shoring without rock bolts when designing and constructing the facility.

One of the purposes of this in-situ experiment was to obtain scientific basis for the development of a quality control methodology when all sections are sprayed in the future geological disposal facility. The results of planned laboratory testing during backfilling by spraying and preliminary tests on the ground surface indicate that the applied quality control methodology was highly evaluated and appropriate for the quality control of backfill material as described above. The bulk density measurement with the 3D scanner tended to be slightly smaller than the values obtained by Lab. testing. However, the measurement method has the several advantages such as a certain volume can be calculated in about 20 minutes with no damage of backfilled area and worker’s safety. These are important issues for the quality control related to backfilling of long tunnels focused on the actual geological disposal projects and the results suggested that the quantitative estimation of bulk density should be made calibration curve between the Lab. testing and 3D scanning.

As regards the monitoring, in the view of quality control, it was difficult to conclude in this experiment due to its small swelling pressure, allowance of outflow from backfilled part and in-situ experiment duration. However, the results showed changes of state in backfilled parts, which is important information to understand the real phenomenon in the area and to support the quality of backfilling.

Finally, measurement of the erosion rate of bentonite in backfilled material was performed. The method was simple. The results suggested that it will be roughly predicted the amount of erosion of Kunigel V1 in in-situ with time. It is also valuable result for the design of the backfilling in actual geological disposal.

This study was performed as part of a commissioned project funded by the Ministry of Economy, Trade and Industry of Japan.

REFERENCES

[1] JNC, H17:Development and management of the technical knowledge base for the geological disposal of HLW, JNC TN1400 2005-022, 2005.

[2] Mikake, S., Iwatsuki, T., Matsui, H. & Sasao, S., Current status and next five-year plan of R&D activities of Mizunami Underground Research Laboratory, *Proceedings of DOPAS Seminar*, 2016.

[3] Ohsawa, H. et al., Current status of R&D activities and future plan of Mizunami Underground Research Laboratory. *Proceedings of 2015 International High-Level Radioactive Waste Management Conference (IHLRWM 2015)* (CD-ROM), pp. 371–378, 2015.

[4] JAEA, Planning for in-situ backfilling test to the gallery in the Mizunami Underground Research Laboratory, JAEA Technology 2018-017, 2018. (In Japanese with English abstract.)

[5] NUMO, www.numo.or.jp/topics/safetycase_setsumei4.pdf. Accessed on: Aug. 2018. (In Japanese.)

[6] JNC, H17: Development and management of the technical knowledge base for the geological disposal of HLW -Repository engineering technology, JNC TN1400 2005-015, pp. 32–35, 2005. (In Japanese.)

[7] SKB, Deep repository-engineered barrier system. Piping and erosion in tunnel backfill, SKB P-06-72, 2008.

HAZARDOUS WASTE MANAGEMENT IN COSTA RICA: AN ACADEMIC–SMALL COMPANY COLLABORATION

LUIS G. ROMERO-ESQUIVEL & ANDREA ACUÑA-PIEDRA
Environmental Protection Research Center (CIPA), School of Chemistry,
Instituto Tecnológico de Costa Rica (ITCR), Costa Rica

ABSTRACT

Hazardous waste treatment options in developing countries are limited. The main problems include scarce or no specialized facilities, and a lack of analytical tools for characterization. In some cases, there are also minimal regulations. Costa Rican hazardous waste legislation dating from the 1990s was improved over the last decade. We describe the management system implemented in Costa Rica that permits an increase in the amount of hazardous waste that could be properly managed, from around 7,432 tons in 2015 to 31,268 tons (76%) in 2019. Similarly, during the same period, the number of official waste-generator companies and specialized waste management companies increased from 107 to 736 (85%), and from 11 to 35 (69%), respectively. We also present the collaboration of a public university with a local small company. Characterization of hazardous waste and evaluation of possible treatment methods were performed at a laboratory level in the university. Later, the small company, under university staff supervision, scaled up the selected treatment method. This small company has two reactors with capacities for 2,000 and 500 kg, both are equipped with pH and oxidation potential reduction sensors that control the pumps used for adding acid, base or oxidants into each reactor. Examples of successfully treated liquid waste are sulfide and cyanide wastes from a metal finishing company. Both types of wastes were oxidized using chlorine, to obtain calcium sulfate and nitrogen gas, respectively. Heavy metal waste is precipitated and then solidified using Portland cement. Hazardous organic waste is normally sent to a local cement kiln for co-processing; however, some waste does need pre-treatment before burning, such as the case of di-isocyanates, which, during burning, could generate hydrogen cyanide. So the waste has to be transformed into urethanes by an aqueous reaction, with ethanol and ammonia. This study demonstrates cooperation between academia, through technical knowledge and analytical skills, and industry, providing appropriate infrastructure and management in accordance with local regulation.

Keywords: Costa Rica, developing countries, environmental legislation, hazardous waste, hazardous waste treatment, university–industry cooperation, waste disposal methods.

1 INTRODUCTION

The use of chemicals in industry, commerce and global research is becoming more common. The chemical industry is expected to continue to grow steadily until 2030 [1]. The mishandling of chemicals, as well as their residues, can adversely affect human health and the environment. The World Health Organization (WHO) estimates that at least 355,000 people die from accidental poisoning each year, with two-thirds of those deaths occurring in developing countries. Poisoning is closely related to excessive exposure to toxic chemicals, including pesticides, and the inappropriate use of these products [1].

As part of the use of chemicals, there is an intrinsic generation of hazardous waste. In response to the identification of global problems regarding treatment and management of hazardous waste at the international level, the "Basel Convention on the Control of Transboundary Movements of Hazardous Wastes and their Disposal" was adopted. This agreement establishes, among other measures, the possible treatments that can be offered to process hazardous waste, for either its recovery or the reduction of risk. As a signatory to this agreement since 1994, Costa Rica subscribes to the treatment options that are allowed for hazardous waste mentioned in that convention [2]; however, only those that may lead to

WIT Transactions on Ecology and the Environment, Vol 247, © 2020 WIT Press
www.witpress.com, ISSN 1743-3541 (on-line)
doi:10.2495/WM200151

recycling, recovery, regeneration and reuse have been considered appropriate, among which are [3]:

- Use as fuel (other than direct incineration) and power generation.
- Recovery or regeneration of solvents.
- Recycling or recovery of organic substances (that are not used as solvents); and metals, metallic compounds and other inorganic materials.
- Regeneration of acids, bases and used oils.
- Recovery of components from catalysts and/or components used to reduce contamination.
- Soil treatments that benefit ecological agriculture or other improvement.

As part of the fulfillment of this agreement, as well as the interest of Costa Rica in preserving its environment and the health of its inhabitants, legislation was created surrounding the integrated management of waste, with the proper management of hazardous waste being accorded great importance; however, compliance with that legislation can be complex, especially for small businesses that work as waste managers. Technologies and methods developed to treat and/or recover energy from waste are often complex; in addition to limitations on, or unavailability of, the necessary resources and knowledge. These small companies has seen a need to seek support from the academy, which, through their experience and knowledge, could provide the necessary tools.

This article describes the Costa Rican legislation and procedures that must be followed in order to responsibly manage hazardous chemical waste. In addition, we present the specific case of successful synergy between a small waste management company, Greco Chemical; and the public university, Costa Rica Institute of Technology (ITCR).

2 HAZARDOUS WASTE MANAGEMENT IN COSTA RICA

2.1 Legal context

In Costa Rica, the Ministry of Health has the responsibility for acting as the lead institution in integrated waste management. This institution is responsible, among other things, for issuing regulations and ensuring compliance; defining indicators in this area; encouraging inter-institutional public environmental action; and encouraging the development of new technologies, models and investments to improve waste management [4]. This ministry is also supported by the Ministry of Environment and Energy (MINAE), in actions related to the management of hazardous waste, such as the creation of regulations and tools, verification of compliance, updating, definition of goals and indicators, development of diagnoses on the national situation regarding hazardous waste, carrying out environmental impact assessments and fulfilling the functions of the Environmental Comptroller, for companies that manage hazardous waste [2]. MINAE helps to promote socially fair and environmentally appropriate commercialization structures for recoverable waste and the products obtained from this, as well as for requesting life cycle analyses or similar studies to evaluate risks and impact [4].

The implementation of regulations dates back to 1998, when three regulations, remaining in force to date, were decreed. Initially, Regulation 27000-MINAE, "Regulation on the characteristics and list of industrial hazardous waste" was created. This regulation establishes the characteristics that allow waste to be classified as hazardous. Within these, are mentioned the limiting parameters for infectiousness, toxicity, flammability, explosiveness, corrosion and reactivity [5]. Also decreed, Regulation number 27001-MINAE, "Regulation for the management of industrial hazardous waste" [6]. It establishes eight stages and guidelines to be met by hazardous waste generators and managers at each step. The stages are: generation, accumulation, pretreatment, storage, transportation (outside the generator facilities), treatment, energy recovery and final disposal [6]. It is understood that some waste does not go through all the stages; as, for example, conditioning within the generator facilities is not always possible or necessary. Moreover, not all waste can be assessed, some are only treated and sent for final disposal with a lowered degree of risk. Regulation 27002-MINAE was created: "Regulation on the procedure to carry out the extraction examination to determine characteristics that make a waste hazardous due to its toxicity to the environment" [7], with the objective to determine whether a residue has hazardous characteristics, using a leaching procedure similar to the US Environmental Protection Agency (USEPA) toxicity characteristics leaching procedure (TCLP) [8]. Finally, there is regulation 24715-MOPT-MEIC-S, "Regulation for the land transport of dangerous products", rules on the transport of hazardous materials (not only waste) [9].

In 2010, the country's first "Law for Waste Management" (Law 8839) was created in Costa Rica. In 2012, the regulations of this law (37567-S-MINAE-H): "General regulation of the law for waste management" came into force. Both documents focus not only on hazardous waste, but also on ordinary waste as well as waste classified as "Special". Among the topics included in this law and its regulations, is the need for the companies responsible for waste handling, at any stage, to ensure compliance with proper management and traceability. The responsibilities of the waste generators and managers are identified, as well as those of the government entities related to compliance, with regulations on this matter. The sanctions that must be imposed for poor waste management are stipulated, in addition to all the requirements and restrictions related to transboundary movements. For example, in the specific case of hazardous waste, transportation without authorization, leaving hazardous waste in unauthorized places and the mixing of hazardous waste with non-hazardous waste are penalized in a "very serious" manner [4]. Carrying out any of these activities may be punished with a fine of approximately ($16,000–$79,000), in addition to the repair of any environmental damage.

From 2010, when the Law for Waste Management was created, a series of modifications was made to the legislation, regarding the management of hazardous waste, especially because Costa Rica made the decision to become part of the Organization for Economic Cooperation and Development (OECD). Thus, waste classified as hazardous was redefined according to the list in the "Basel Convention on the Control of Transboundary Movements of Hazardous Wastes and their Disposal": the waste became subject to control by the OECD, in Decision C (2001)107. The responsibilities of waste generators and managers in all stages of waste management are specified in greater detail. Therefore, modifications were made to the legislation, and regulation number 41052-S: "Regulation of recoverable waste recovery centers" was introduced in June 2018 [10]. Here, compliance regarding physical space, good practices and documentation showing that waste managers have carried out storage, treatment, recovery or final disposal of waste (including those of a hazardous type) are indicated. Additionally, in January 2019, regulation 41527-S-MINAE: "General regulation for the classification and management of hazardous waste" came into force [2]. This last regulation expands the necessary requirements for the integrated management of hazardous waste in Costa Rica.

2.2 Requirements for the management of hazardous waste in Costa Rica (generators and waste managers)

Considering all the existing regulations before the Law 8839 of 2010, as well as the regulations that arose after this, both waste generators and waste managers are forced to comply with a large number of requirements [11], such as:

- Carry out integrated management of the waste generated (at all stages, from generation to final disposal).
- Implement an Integrated Waste Management Program. This is a document that contains the diagnosis of the current situation regarding the waste management of a generator or a waste manager, in addition to proposing improvements. This document must have a section dedicated exclusively to the integrated management of hazardous waste. The document details each hazardous waste that is generated or managed, including its destination, risk of handling, as well as ways to prevent and reduce its generation and to minimize risk.
- Register with the Ministry of Health any movements and release of hazardous waste.
- Notify the MINAE Environmental Comptroller of the dispatch of waste outside generator facilities.
- Have the necessary permits for the transboundary movement of hazardous waste.
- Guarantee that the packing and labeling of hazardous waste is carried out in accordance with the aforementioned regulations.
- Have adequate storage, recovery, transportation, treatment or final disposal facilities, as dictated by regulations.
- Immediately inform the Ministry of Health in the case of any disappearance, loss or spillage of hazardous waste.
- Prepare and implement an Occupational Health Plan and an Emergency Care Plan.

Meanwhile, in the accumulation stage, it is requested that separation points be placed as close as strategically possible to the waste generation focus. Chemical incompatibilities must be respected; and there must be appropriate containers, labeled according to the type of waste to be accumulated.

For the storage stage, the same requirements for safety and labeling of containers, and respect for incompatibilities during accumulation, must be followed. Also, waste must be taken outside the generator facilities for assessment, treatment or final disposal at least once a year, for each type of waste contained, or whenever storage reaches 1,000 L. Care must be taken in storage, with pallets also according to the type of waste.

Regarding transportation, the waste manager and the generator who contracts transport must comply with the rules stated for transporting hazardous materials. Only trained drivers and authorized transportation vehicles can perform this task. They must respect chemical incompatibilities, with pallets and safe packaging. They must carry documentation indicating the waste(s) being transported, with notification of that movement made to MINAE and with national circulation permits.

Included among the requirements that waste managers must abide by for waste treatment are environmental impact studies, risk analyses, and use of treatment methods that lead to the recovery of the waste or at least to reduce its dangerousness, in the event that it cannot be recovered.

The final disposal allowed in Costa Rica is the export or shipment to sanitary landfills, as long as it is waste that has been treated or placed in safety cells in the sanitary landfill, cells that prevent the waste from causing damage.

It is very important to mention that Costa Rica's regulations maintain the concept of "survival of responsibility", which is defined as: "The entire responsibility of the generator remains until the hazardous waste is assessed or disposed of permanently. The waste manager hired by the generator will be jointly responsible for it, once the waste has been received for its integrated management" [2].

2.2.1 Notifications of waste movements

As mentioned, waste must be take outside of the generator facilities at least once a year (6 months, in the case of the waste managers), or when there is a stored quantity of 1,000 L. Often, both due to space availability and risk issues in storage, waste must be removed much more frequently, making the transportation of waste a daily practice throughout the national territory.

Each of these movements must be notified to the MINAE Environmental Comptroller. In order to carry out the notifications effectively, the System of Integral Management of Hazardous Waste (SIGREP) has been created by this ministry. This is a web platform that allows generators to register all of the hazardous waste they generate; also, waste managers can register the hazardous waste that they can accept and for which they can provide transportation, storage, treatment, recovery or final disposal. In addition, this platform enables the registration of drivers and vehicles that have the corresponding training and permits to carry out the transportation work. Prior to each movement of waste outside the facilities of the generator or the waste manager, MINAE is notified of the waste being transported, its physical and chemical characteristics, incompatibilities, quantity, destination, the driver's name and vehicle in which it is transported. Fig. 1 graphically presents the general process of hazardous waste management in the country.

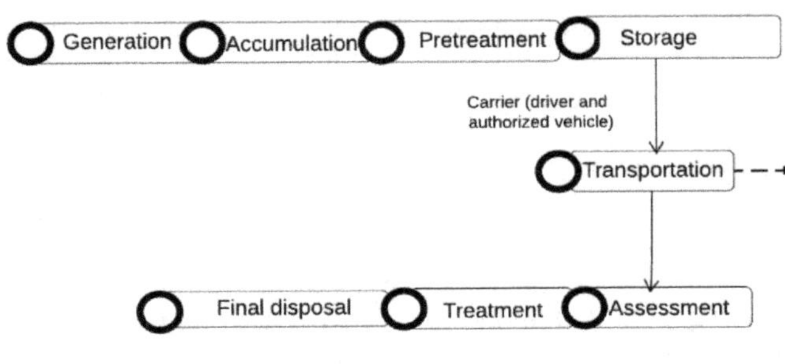

Figure 1: General process of hazardous waste management in Costa Rica.

3 APPLICABILITY OF INTEGRATED HAZARDOUS WASTE MANAGEMENT IN PRACTICE

Although the legislation is exhaustive, it is not always possible to apply it as the regulations suggest. Currently, many waste managers work illegally, and many waste generators do not meet the minimum requirements established; however, the SIGREP platform has provided data that supports the improvements made and also provides evidence for decision-making.

Subsequently, we present the results reported on the platform, followed by details regarding an authorized manager.

3.1 Hazardous waste management from 2015 to 2019

For the year 2019 [11], the Costa Rica Environmental Comptroller reports on the management of 31,268 tons of hazardous waste, transported from the generation point to the different waste managers. The processes these residues were subjected to were: use as an alternative fuel, solvent recovery, recovery of organic substances, neutralization, oil regeneration, and recovery of lead.

The waste managed in the largest quantity was lubricating oils (9,218 tons); followed by solids contaminated with harmful impurities (4,400 tons); and then third, paint, solvent and sludge residues produced by painting, coating and cleaning activities (4,300 tons). Other waste managed in lesser quantities were: contaminated fuels, plastic waste, contaminated absorbent materials and textiles, lead acid batteries, solvents, and other oils.

Since the introduction of this platform, there has been an increase in the amount of waste processed and the number of generating companies, managers, transporters and transport manifests reported (Table 1) [11]. This indicates the impact that the new legislation has had on the entire waste chain, from generation to final disposal.

Table 1: Hazardous waste generation and management in Costa Rica from 2015 to 2019.

Topic	2015	2019	Increase (%)
Total hazardous waste (ton)	7,439	31,268	76.2
Generating companies	107	736	85.5
Management companies	11	35	68.6
Transport companies	23	32	28.1
Total transportation manifests	1,192	12,138	90.2

3.2 Example of a small hazardous-waste management company

Greco Chemical is a small Costa Rican company with which the Research Center for the Protection of the Environment (CIPA) of the ITCR has had a successful relationship regarding the research and development of efficient methods for treating hazardous chemical waste, since 2012. Tests were initially carried out in the ITCR laboratories, to design an appropriate treatment method for a series of hazardous wastes (detailed in next section). Once these methods were validated, they were able to be scaled up in the Greco Chemical facility.

Currently, the company has a 200 m^2 treatment and storage facility and a small laboratory to characterize the waste and to test treatment methods. The plant has the capacity to treat waste in small quantities (25 mL vials), up to 20,000 kg. It has two reactors with capacities of 2,000 kg and 500 kg, with pH and oxidation potential reduction (ORP) controllers for dosage of chemical products.

As a waste manager, the company has authorization from the Ministry of Health and is registered on the SIGREP platform, which, as previously mentioned, allows for the notification of waste movements from generators to management companies.

Currently, the company offers hazardous waste treatment services by both physical and chemical methods (Table 2, Fig. 2). In the case of physical treatments, it offers filtration, gravity separation, evaporation, carbon absorption, ion exchange and solidification (immobilization). The chemical treatments carried out by Greco Chemical are: neutralization, reduction, oxidation-reduction, oxidation decomposition, substitution, modification or decomposition of functional groups. In addition, the company provides treatment to waste to

Table 2: Hazardous waste management by Greco Chemical.

Waste type	Origin	Treatment	Final disposal	Annual amount processed (tons)
Acid solutions	Electroplating Teaching lab waste Expired chemical reagents	Neutralization	Co-processing	15
Basic solutions	Electroplating Teaching lab waste Expired chemical reagents	Neutralization	Co-processing	15
Organic solvents	Pharmaceutical industry Teaching lab waste Expired chemical reagents Paint industry	Some are transferred to a recycling plant for production of thinner, e.g. xylene. Other solvents undergo no treatment: they are only used to co-process.	Recycling Co-processing	8
Cyanide waste	Jewelry industry	Deactivation by ORP, using sodium hypochlorite	Wastewater goes to co-processing Silver chloride solid is reused	0.630
Ink waste	Lithographic industry	pH is checked	Co-processing	3
Solid hydrocarbon-contaminated cloth or plastic	Mechanical workshop	Trituration	Co-processing	10
Liquid waste with heavy metals	Laboratory waste	Precipitation by chemical reaction, controlling pH	Heavy metals encapsulated, taken to a landfill TCLP test	0.600

be managed through co-processing in a cement kiln. In the case of waste that cannot be recovered, it is disposed of in a sanitary landfill.

At Greco Chemical, prior to submitting any waste to treatment, it is analyzed, in order to determine the best way to manage it; and always seeking recovery, rather than co-processing, and trying to avoid final disposal in sanitary landfills. Fig. 2 shows the decision path followed in the company.

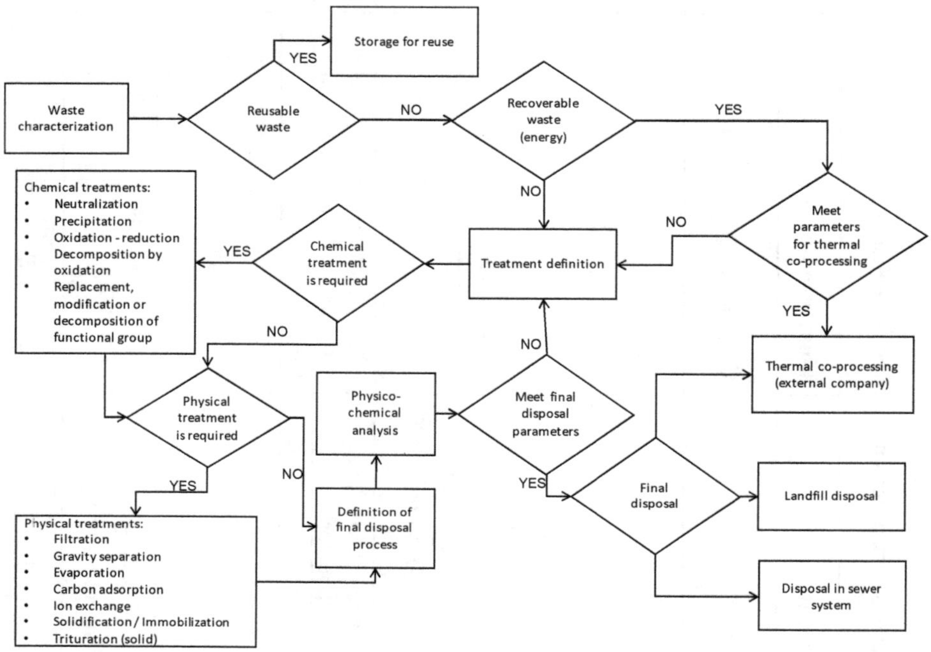

Figure 2: Decision diagram for waste management.

Among the chemical treatments carried out is the treatment for chromium residues (eqns (1) and (2)).

$$2H_2CrO_4 + 3Na_2HSO_3 + 3H_2SO_4 \longrightarrow Cr_2(SO_4)_3 + 3NaHSO_4 + 5H_2O, \quad (1)$$

pH = 2.5–3.0, ORP = 250–300 mV,

$$Cr_2(SO_4)_3 + 3Ca(OH)_2 \longrightarrow 2Cr(OH)_3 + 3CaSO_4, \quad (2)$$

pH = 8–9.

Cyanide waste used in the jewelry industry is treated by oxidation with sodium hypochlorite (eqns (3) and (4)).

$$NaCN + NaOCl \longrightarrow NaCNO + NaCl, \quad (3)$$

pH more than 10; ORP = 300–350 mV,

$$2NaCNO + 3NaOCl \longrightarrow 2CO_2 + N_2 + 2NaOH + 3NaCl, \quad (4)$$

pH = 7.5–8.0; ORP more than 650 mV.

Another example is the destruction of isocyanates by reaction with alcohol in an alkaline medium, to produce urethane (eqns (5) and (6)) [12], which is subsequently co-processed by the cement industry

$$ROH + NH_3. \longrightarrow RO^- + NH_4^+, \quad (5)$$

$$R^1\text{-}N{=}C{=}O + {}^-OR. \longrightarrow R1\text{-}NH\text{-}COOR + NH_3. \quad (6)$$

3.3 Final remarks: Importance of the synergy between the university and small hazardous waste managers

In Costa Rica, scientific research is mainly carried out in public universities; at the same time, synergy with private companies permits optimal results for technological and industrial development. In the particular case of Greco Chemical and the academic ITCR, a synergy occurred due to the educational center's need to responsibly manage the waste from its laboratories, while Greco Chemical lacked the knowledge and equipment of the specialized research laboratories; therefore, the ITCR provided the support required by Greco Chemical. As a result, it was possible to develop treatment systems and appropriate logistics for treatment and disposal of hazardous waste that enabled the environmental performance within the country to be enhanced.

ACKNOWLEDGEMENTS

The authors thank both Greco Chemical, for the information provided about the quantity of materials processed and its treatment system; and the Scholarship Committee of the ITCR, for registration fee payment.

REFERENCES

[1] Strategic Approach to International Chemicals Management (SAICM), Informe de la Conferencia Internacional sobre Gestión de los Productos Químicos relativo a la labor realizada en su tercer período de sesiones [Report from the *SAICM International Conference*, presented at the third period of sessions], Nairobi, 2012. www.who.int/ipcs/capacity_building/chemicals_management/saicm_iccm3_sp.pdf?ua=1. Accessed on: 12 Mar. 2020.

[2] Asamblea Legislativa de Costa Rica [Costa Rica Legislative Assembly], Reglamento general para la clasificación y manejo de residuos peligrosos (41527-S-MINAE) [General rules for classification and handling of dangerous residues (41527-S-MINAE)], Costa Rica, 2019. www.pgrweb.go.cr/scij/Busqueda/Normativa/Normas/nrm_texto_completo.aspx?param1=NRTC&nValor1=1&nValor2=88120&nValor3=114959&strTipM=TC. Accessed on: 2 Mar. 2020.

[3] Basel Convention, On the control of transboundary movements of hazardous wastes and their disposal. *Basel Convention*, Basel, Switzerland, 1989. www.basel.int/Portals/4/Basel%20Convention/docs/text/BaselConventionText-e.pdf. Accessed on: 12 Mar. 2020.

[4] Asamblea Legislativa de Costa Rica [Costa Rica Legislative Assembly], Ley para la gestión integral de residuos (8839) [8839 Law about complete management of residues], Costa Rica, 2010. www.pgrweb.go.cr/scij/Busqueda/Normativa/Normas/nrm_texto_completo.aspx?nValor1=1&nValor2=68300. Accessed on: 2 Mar. 2020.

[5] Asamblea Legislativa Costa Rica [Costa Rica Legislative Assembly], Reglamento sobre las características y listado de los desechos peligrosos industriales (27000-MINAE) [27000-MINAE Rule about the characteristics and listing of dangerous industrial wastes], Costa Rica, 1998. www.pgrweb.go.cr/scij/Busqueda/Normativa/Normas/nrm_texto_completo.aspx?param1=NRTC&nValor1=1&nValor2=44467&nValor3=95389&strTipM=TC. Accessed on: 2 Mar. 2020.

[6] Asamblea Legislativa Costa Rica [Costa Rica Legislative Assembly], Reglamento para el Manejo de los Desechos Peligrosos Industriales (27001-S-MINAE) [27001-S-MINAE Rule about the handling of dangerous industrial wastes], Costa Rica, 1998. www.pgrweb.go.cr/scij/Busqueda/Normativa/Normas/nrm_texto_completo.aspx?param1=NRTC&nValor1=1&nValor2=54836&nValor3=114966&strTipM=TC. Accessed on: 2 Mar. 2020.

[7] Asamblea Legislativa Costa Rica [Costa Rica Legislative Assembly], Reglamento sobre el procedimiento para llevar a cabo la prueba de extracción para determinar constituyentes que hacen un residuo peligroso por su toxisidad al ambiente (27002-MINAE) [27002-MINAE Rule about the procedures for testing the extraction to determine the components that make a dangerous waste due to its environmental toxicity], Costa Rica, 1998. www.pgrweb.go.cr/scij/Busqueda/Normativa/Normas/nrm_texto_completo.aspx?param1=NRTC&nValor1=1&nValor2=54838&nValor3=60088&strTipM=TC. Accessed on: 2 Mar. 2020.

[8] United States Environmental Protection Agency (US EPA), *Test Methods for Evaluating Solid Waste, Physical/Chemical Methods, SW-846, Method 1311*, US Government Printing Office: Washington, DC, 1992.

[9] Asamblea Legislativa Costa Rica [Costa Rica Legislative Assembly], Reglamento para el transporte terrestre de productos peligrosos (24715-MOPT-MEIC-S) [24715-MOPT-MEIC-S Rule for the terrestrial transportation of dangerous products], Costa Rica, 1995. www.pgrweb.go.cr/scij/Busqueda/Normativa/Normas/nrm_texto_completo.aspx?param1=NRTC&nValor1=1&nValor2=40953&nValor3=80233&strTipM=TC. Accessed on: 2 Mar. 2020.

[10] Asamblea Legislativa Costa Rica [Costa Rica Legislative Assembly], Reglamento de Centros de Recuperación de Residuos Valorizables (41052-S) [41052-S Rules for the centers for recuperation of valuable residues], Costa Rica, 2018. www.pgrweb.go.cr/scij/Busqueda/Normativa/Normas/nrm_texto_completo.aspx?param1=NRTC&nValor1=1&nValor2=86628&nValor3=112489&strTipM=TC. Accessed on: 2 Mar. 2020.

[11] Contraloría Ambiental [Environmental Comptroller], Informe de la gestion de residuos peligrosos año 2019, segun el SIGREP [SIGREP report on management of hazardous residues], Costa Rica, 2019. www.contraloriaambiental.go.cr/pag/inicio.php. Accessed on 11 Mar. 2020.

[12] Das Bakul, T.V., Reddy, R., Balaji, M.T.K. & Veera Reddy, T., Development of eco-friendly neutralizing agents for toluene di-isocyanate. *Research Journal of Chemical Sciences*, **3**(5), pp. 7–11, 2013.

DISPOSAL OR TREATMENT: FUTURE CONSIDERATIONS FOR SOLID WASTE FROM THE CONSTRUCTION AND DEMOLITION INDUSTRY

SHANNON L. WALLIS[1,2], CHARLES LEMCKERT[1], ROBYN HARDY[1] & TERRI-ANN BERRY[2]
[1]University of Canberra, Australia
[2]Unitec Institute of Technology, New Zealand

ABSTRACT

Each year more than 2 billion tonnes of municipal solid waste are produced globally. The greatest worldwide users of resources in terms of raw materials and energy are from the construction sector. Construction, demolition and excavation waste accounts for a large share of municipal waste and the majority of this waste goes to landfill. Worldwide, the proportion of landfilled construction, demolition and excavation waste compared with the total amount of waste is variable from 13% to 60%+. Many developed countries are now facing issues with land capacity and are less flexible in their approach to solid waste treatment. New Zealand is a relatively young and geographically isolated country with enough available land to be able to trial new options for waste treatment and currently has seven different types of landfills, including construction and demolition and cleanfill options. Biodegradable substances can be treated via processes such as composting and anaerobic digestion, but substances which are either hazardous or inorganic in structure are generally considered to be untreatable and therefore reduce practical options to landfill or incineration. As any potential treatment in landfills is limited by less than optimal environmental conditions, their primary purpose is simply to hold and isolate waste. Incineration has high energy costs and does not support a low carbon economy. As neither of these options present an ideal long-term solution to solve this waste problem, this paper will consider sustainable options for the disposal and treatment of construction, demolition and excavation waste with a focus on hazardous materials. It will consider New Zealand as a potential case study location for trialling solid waste treatment options whilst discussing waste issues in other countries such as Australia, the United Kingdom and the United States of America. It will also identify barriers which may prevent the treatment of solid waste including considerations for protecting public health and safety.

Keywords: waste management, waste disposal, waste treatment, CD&E waste, hazardous waste.

1 INTRODUCTION

Worldwide, more than two billion tonnes of municipal solid waste (MSW) are produced annually [1], where the construction sector is the largest global user of raw material and energy resources [2]. Construction, demolition and excavation (CD&E) waste contributes to a large share of the total MSW in developed countries [2]. Typically, construction and demolition (C&D) wastes are defined by activity [3] and consist of relatively inert materials such as concrete and bricks.

The average proportion of total waste reaching landfills which is derived from CD&E waste is 30% [4] although this proportion varies globally from 13% [5] to 60%+ [6], [7]. In 2016, 61% of waste generated in the UK was classified as CD&E waste [6] whereas Australia reported 31% CD&E waste [8] (Table 1). In Hungary, 23% of waste was labelled as CD&E [9] whereas Romania reported no CD&E waste [9]. This could be due to individual countries either not reporting their figures or simply not having the separate CD&E waste stream but does demonstrate a lack of certainty around this data globally. Although the United Nations Environmental Programme (UNEP) [2] reported that the bulk of waste disposed via landfills

WIT Transactions on Ecology and the Environment, Vol 247, © 2020 WIT Press
www.witpress.com, ISSN 1743-3541 (on-line)
doi:10.2495/WM200161

Table 1: Waste generation and breakdown of waste in seven countries [5]–[10].

Country	Year	Total mass	Waste generation per capita (2015)	CD&E	Household	C&I**	Others
UK [6]	2016	222.9	1.8	61%	12%	19%	8%
Australia [8]	2016	66.8	2.2	31%	21%	31%	18%
Germany [7]	2015	351.2	2.1	60%	15%	17%	9%
Japan [10]	2011	380	–	20%	11%*	63%	6%
South Africa [5]	2017	42.7	2	13%	20%	7%	60%
Hungary [9]	2016	16	–	23%	18%	34%	25%
Romania [9]	2016	178	–	0%	2%	95%	3%

Notes: Total and waste generation per capita as million tonnes. *Food, paper products and communication devices only. **C&I include commercial and industrial, as well as agricultural and energy generation sectors.

was derived from CD&E waste in developing countries, this has not been supported by other research (as demonstrated in Table 1).

Regardless of the reliability of the CD&E waste data, the vast scale of the problem is concerning and other options to landfill disposal need to be investigated. Established waste disposal and treatment processes for MSW include composting, incineration, landfill and recycling, globally, technology choice is based predominantly on land space availability. In terms of CD&E waste, previous research indicates a high potential for recovery [5], [7], [8] (with the highest rate available in the absence of hazardous waste (UK) [6]). In New Zealand (NZ), it is possible to recycle and/or reuse most streams of C&D waste. For example, Green Gorilla, a waste management company in NZ recycles timber, non-ferrous metals, cardboard, plasterboard, steel, rocks and concrete. However, there are other barriers preventing effective recycling or reuse which include a lack of training and education to encourage on-site waste separation, space limitations on site for adequate materials storage and a clear lack of incentives to reduce waste. For hazardous CD&E derived wastes, recycling or reusing waste is rarely if ever a viable option. So, despite impressive improvements in the capability to be able to divert this waste stream from landfill, this disposal route remains the most common option, globally.

The NZ government defines hazardous waste as that which "contains hazardous substances at sufficient concentrations to exceed the minimum degrees of hazard specified by Hazardous Substances Regulations 2000" or meets the definitions for infectious substances or radioactive materials [11]. The hazardous characteristics that form the basis for these criteria include explosiveness, flammability, capacity to oxidise, toxicity, corrosiveness or eco-toxicity [11]. The lack of hazardous waste management in NZ was identified as a key issue twenty years ago [12] and yet the NZ Waste Strategy (2010) gives little or no mention to hazardous waste disposal [13]. Furthermore, there has been little focus on hazardous wastes which are created due to past activities such as the large quantities of contaminated land which currently follow a similar hazardous waste disposal protocol.

Globally, we produce 400 million tonnes of hazardous waste each year [14]. However hazardous waste from the CD&E sector consists of three distinct types which should be considered separately. These include, hazardous products (HP) used during construction, e.g. paints and solvents, which are often in liquid form. Hazardous waste (HW) by which we are referring to hazardous components of building products for example, fluorescent lamps containing mercury or electrical components containing polychlorinated biphenyls (PCBs).

Finally, hazardous by-products (HBP) which are often solid wastes containing a low concentration of hazardous component, such as asbestos or lead contaminated soils. In fact, contaminated soil waste and asbestos waste contributes a significant portion of low concentration, high volume waste to landfill. A previous study quantified the C&D waste materials in all states across Australia (2008–2009) where 1,055,797 tonnes of contaminated soil waste and 728,477 tonnes of asbestos waste were generated [15].

Lack of available land is becoming an increasing issue for many developed countries reducing flexibility in terms of future solid waste treatment. Are there other options for the disposal of CD&E waste? And, do these options include more sustainable waste treatment? This paper will investigate the current options available whilst highlighting the issues associated with long-term landfill disposal of hazardous substances.

2 LANDFILL DISPOSAL

Modern landfills are designed to minimise the impact of MSW on the environment and human health, providing storage but requiring land-use restrictions and continuing maintenance [16]. Landfills are the most common method of waste disposal in NZ, with an estimated 3.2 million tonnes of waste to MSW landfills in 2006 [17], [18]. This waste typically includes 28% organics, 16% rubble, 11% timber, 8% plastics, 7% paper and less than 5% of glass, metals, and textiles [19]. In NZ, there are currently seven different types of landfills where waste can be disposed; five non-hazardous landfills (MSW, managed, C&D, cleanfill, industrial) [20] and two hazardous landfills (Class A, Class B) [21]. Cleanfills are defined in NZ as, "material that when buried will have no adverse effect on people or the environment" (e.g. clay, brick, concrete etc.) [20].

From 1995 to 2007, the number of landfills in NZ reduced from 327 to 60, (often as a result of less than optimal environmental controls), with just 54% utilising engineered liners for leachate containment [18]. Nevertheless, the amount of waste to landfill in NZ appears to be increasing, rising to just over 1 tonne/capita/yr [22], representing close to 5 million tonnes of waste per annum. It has been estimated that 26% of waste to landfill is derived from C&D waste [23]. However, this value does not include waste that is sent to cleanfill sites which was estimated to be between 2.7 and 3.7 million tonnes in 2007 [23].

For hazardous substances, Class A landfills have engineered systems designed to the Centre for Advanced Engineering's Landfill Guidelines (2000), in comparison, Class B landfills are existing landfills that do not meet this guideline, and have little to no engineered systems [21]. Despite landfills specific for hazardous waste, it is estimated that approximately 14% of waste to four MSW landfills was deemed potentially hazardous [19].

Although there is currently a heavy reliance on waste disposal via landfill, NZ is a relatively young country with a small population and could investigate alternative options for the treatment of CD&E and associated hazardous wastes. This would be advantageous to support its reputation as an environmentally focused society and to provide a more sustainable option to protect future generations and a fast-growing population.

3 THE FUTURE OF LANDFILL TECHNOLOGY

There is evidence to suggest that the world has reached a level of stagnation with respect to landfill design and waste disposal. In 1995, a seminar in the United States was called to examine "an alternative approach to landfill operation that greatly minimizes long-term risks associated with potential landfill containment system failure" [24]. This alternative approach was bioreactor landfilling and it was stated that "this method represents the future of waste disposal because it transforms waste disposal practice from a passive system to an active process" [24]. In 2000, NZ revised their Centre for Advanced Engineering Landfill

Engineering Guidelines (1992), now called Landfill Guidelines (2000) [25]. These guidelines stated, "bioreactor and aerobic landfills have not been designed or operated in New Zealand to date", however it was stressed that overseas research and trials should be monitored for their applicability to NZ (e.g. bioreactor landfills) [25]). The Landfill Guidelines (2000) were superseded in 2016 (with a further revision in 2018) to Technical Guidelines for Disposal to Land (August 2018), with no reference to bioreactor landfills [26]. Which shows little progress made in terms of landfill design in 25 years.

Current landfill designs and management have a number of issues including a poor level of degradation of waste and issues associated with leachate toxicity due to increased chemical use [27]. Traditional landfills have a large land requirement, and with a growing global population, land space is at a premium. In addition to this, there is a trend towards increasing disposal costs for hazardous waste disposal, which in turn is encouraging illegal dumping of waste [28]; an issue which has been forewarned for New Zealanders over 35 years previously [29].

Issues with landfill failures are widespread despite apparent engineering developments. In the late 1980s problems began to arise for the community of Oakland County Waterford Township (US) in the form of groundwater contamination [30]. It is unclear whether the failure was a result of leachate collection and removal or liner fault [30]; however, it is clear that at the time of landfill construction, guidelines and regulations were less stringent than today. In developing countries, the regulations regarding landfill design are not as well controlled, which has contributed to many landfill failures (Table 2).

Table 2: Worldwide landfill failures [31]–[38].

Location	Date	Cause	Fatalities
Payatas, Manila, Philippines [31]	Jul. 2000	Heavy rain triggering landslides	More than 200
Leuwigajah, Java, Indonesia [32]	Feb. 2005	Explosion due to sudden biogas release, 3 days of high rainfall	Approx. 143
Morro do Bumba, Niteroi, Brazil [33]	Apr. 2010	Heavy rain triggering floods and landslides	Approx. 200 (across the region)
Baguio, Philippines [34]	Aug. 2011	Typhoon – collapse of retaining wall	5
Guatemala City, Guatemala [35]	Apr. 2016	Heavy rain	Approx. 30
Koshe, Ethiopia [36]	Mar. 2017	Conflicting reasons	Approx. 120
Meethotamulla Garbage Mountain, Sri Lanka [37]	Apr. 2017	Instability – approx. 48.5m high pre-collapse	Approx. 40
Verter Recycling Landfill, Basque Region, Spain [38]	Feb. 2020	Instability – with asbestos waste release	2

In the past, an acceptable disposal route for hazardous or "special" waste was selected to reduce the possibility of future problems arising from the disposal process [29]. At this time, observations from UK landfills informed that "sensible landfill is realistic and an ultra-cautious approach to landfill of hazardous waste is unjustified" [29]. However, we have observed since then that there is an increased risk to the public when landfills fail, as there is always a possibility that hazardous substances could be exposed. Landfill failures can happen at any time and in any country as evidenced by the landfill failure along the Fox River of NZ's West Coast, where asbestos and hazardous substances were suspected to have been among the strewn rubbish [39]. Even if landfills have not failed yet, there is the risk of collapse, such as Kettle Park in Dunedin, NZ, where large swells are impacting and washing away the protective dunes [40]. Along the coasts of England and Wales, 1,000 historic landfills are at risk of failure by erosion, posing a substantial risk to the local populations [41], [42]. This includes the former landfill in East Tilbury that has been leaking waste and hazardous substances into the Thames River [42]. Landfill failures are not limited to waste discharge but can also include gas leaks, such as the methane leak from Sunshine Landfills in Melbourne, Australia [43]. Failures of historic, recently closed, and active landfills is a trend that could become more frequent as landfills are exposed to more extreme weather events combined with poor siting and lack of adequate regulation. Whilst landfills provide public health protection in the short-term, they can still pose a long-term risk made worse by environmental and human-derived change on climate and land-use.

4 TREATMENT OPTIONS

4.1 Chemical and physical processes

Chemical transformation processes have potential to reduce toxicity prior to further treatment or disposal. This type of process is often expensive and by-products may be hazardous themselves and pose a further risk. Chemical methods include complexation, neutralisation and oxidation [44] whereas physical processes include complete thermal destruction (incineration) as well the removal of substances from aqueous solution via processes such as carbon adsorption. For hazardous substances, these may include pre-treatment or pre-disposal stages which can often involve high energy costs. There are a number of other disadvantages associated with these processes including waste production (often toxic), high energy usage and a lack of adaption to wastes of highly variable quality. Similarly, incineration has high energy costs and may not be sustained long-term within a low carbon economy. It is also responsible for the production of air pollutants such as dioxins (unless well-maintained) and highly toxic by-products (such as incinerator ash).

Physical and chemical processes can treat the majority (if not all) of the waste flow and may be useful for pre-treatment however they do not currently provide a sustainable long-term option.

4.2 Biological processes

Biological processes have been slow to adopt partly due to limitations on substances resistant to biodegradation, however recent developments support further investigation. Readily biodegradable substances can be treated via biological processes however substances which are either hazardous or inorganic are generally considered to be untreatable. If waste can be considered to exist along a continuum which describes ease of treatment, where inorganic (e.g. hazardous) substances lie at the extreme end (Fig. 1), there is evidence to suggest that treatment has potential even for the most recalcitrant substances, such as asbestos fibres within a cement matrix [45]. The degradation times for some substances are known and in the case of simple and complex organic molecules, this timescale varies depending upon molecular weight and chemical characteristics such as polarity and microbial toxicity.

The treatment of solid wastes remains problematic however bioremediation (biological process) is capable of treating both solid and liquid wastes. Bioremediation can be either in-

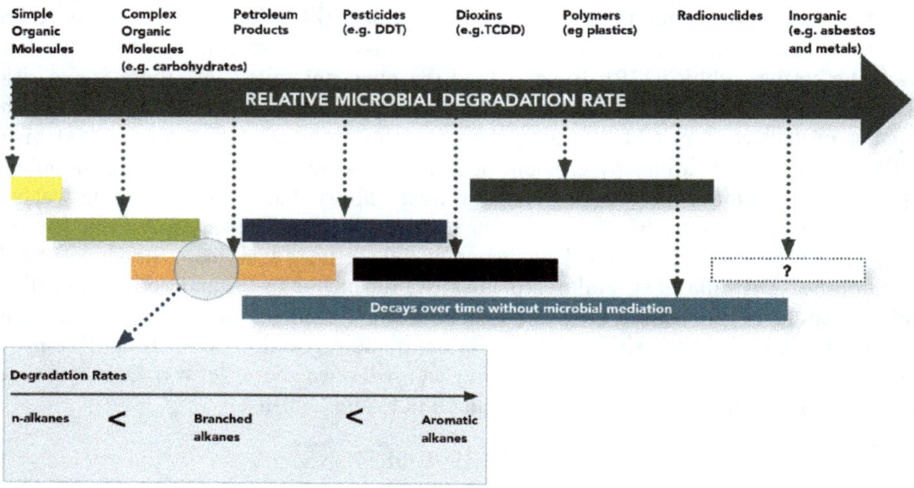

Figure 1: Continuum of relative microbial degradation rate, where length of bar indicates relative degradation timeframe. (Degradation rates of petroleum products [47].)

situ, (bioventing and bioaugmentation); or ex-situ (landfarming and bioreactors (slurry reactors)) [46]. In-situ treatment options are more desirable as they tend to be lower cost and provide less disturbance by treating in place, whereas ex-situ bioremediation involves the excavation of the contaminated soils from the ground and transferred to a separate location for treatment [46]. An example of an ex-situ, partial bioremediation process is the bioreactor (or activated) landfill.

4.2.1 Bioreactor landfills

Bioremediation has been trialled on a limited scale within a landfill-type enclosure, called a bioreactor (or activated) landfill, defined where "liquid or air is injected in a controlled fashion to the waste mass in order to accelerate or enhance biostabilisation of the waste" [44]. These reactors are primarily for organic wastes and are designed and operated under conditions to enhance biodegradation and biogas production. Within bioreactors, ambient moisture levels are critical (as microbial growth is enhanced at moisture contents greater than 40% by weight) so this typically involves the addition of moisture to the process [48]. The addition of moisture serves two purposes, firstly, the creation of conditions favourable for the transport and proliferation of the microbial community and secondly, provides a pathway for mixing organic substrates, nutrients and waste products [48] and the dilution of high concentrations of microbial inhibitors. However, it has been observed that continuously achieving the target moisture content in a landfill is rarely achieved by leachate recirculation alone.

Furthermore, bioreactor landfilling is the least promising option for many wastes (including wood, textiles and paper), based on their estimates of biogas production from anaerobic digestion of organic waste [49]. It appears that currently bioreactor landfills do not deliver the promise of accelerated treatment but still produce undesirable toxic products such as landfill leachate and rank poorly in terms of environmental impact. Due to this and various other limitations mentioned above (such as higher moisture requirements, anaerobic environment, and lack of adequate mixing for transportation of microbes and substrates), the use of bio-landfilling has not served to improve the actual degradation of solid wastes substantially in full-scale applications [48], [50] and has not been adopted in NZ. However, bioremediation on a longer and larger scale may provide a sustainable solution for the treatment of CD&E and associated hazardous wastes, providing public health and safety can be assured.

4.2.2 Bioremediation

Bioremediation utilises living organisms such as plants, microbes and their enzymatic products to reduce toxicity in xenobiotic compounds [51]. Microorganisms are versatile and are capable of rapid adjustment during environmental changes and can therefore serve to protect their ecosystems from deterioration. Microbial-mediated bioremediation is cost effective, sustainable and in-situ application is easily implemented. Either naturally occurring metabolic activity can be utilised during bioremediation for the degradation, transformation or accumulation of many substances [52] or microbial augmentation with non-native species can be effective [51]. Despite the perceived low potential for biological degradation of inorganic species (Fig. 1), successful steps towards bioremediation have been identified recently. For example, the bacterial bioremediation of metal-contaminated waste [51] and the successful use of bioremediation strategies for radionuclides [53], highlighting the potential for bioremediation [54].

Phytoremediation of acidic mining drainage in mine tailings has also shown plants to facilitate the immobilization of other heavy metals such as lead, cadmium, zinc, iron and nickel [55]. Phytocapping, (growth of plants over mine tailings), relies on the addition of an amendment layer on the mine tailings to enable plants to adapt to biotoxins and the acidic pH [55]. Similarly, in the potential bioremediation of other inorganic hazardous materials such as asbestos [56], these and other factors will need to be addressed, especially the lack of organic material and nutrients for growth present within pure asbestos deposits. Unless these can be sufficiently supplemented either by the managed ecosystem itself or the addition of carbon and nutrient rich amendments, bioremediation will be difficult to optimise.

Landfarming, an ex-situ bioremediation technology, is a contained, controlled degradation of contaminants in varied soil conditions to optimize the rate of degradation [57]. Contaminated waste is mixed with soil amendments such as soil bulking agents and/or nutrients then adjusted for parameters such as moisture, pH and aeration (by periodic tilling) [57], [58]. This allows the contaminants to interact with the soil and climate of the site to degrade, immobilize and transform contamination constituents [57]. Landfarming is proven to be most successful in treating petroleum hydrocarbons, and other more chlorinated or nitrated compounds albeit with more difficulty [58]. Potential environmental hazards from landfarming applications can arise from any residual toxic substances remaining within the soil profile; groundwater contamination (leaching); and airborne hazardous substances.

Disadvantages for the use of bioremediation include issues associated with accurate scale-up and a lack of standardised methods for measuring biodegradation rates, resulting in scarce data sets. The use of lab scale experiments to predict field scale results appears risky for determining treatment rates, and bacterial or plant-based processes may produce large volumes of pollutant-loaded biomass which can result in further waste disposal issues. However, the use of fungi for bioremediation processes may present greater opportunities, especially for hazardous substances. In the case of asbestos, the removal of iron from asbestos materials has been indicated using lichens and fungi where the degradation of asbestos using fungi has been tested in controlled laboratory studies [56], [59], [60]. To maximise the potential of bioremediation techniques whilst reducing risk, can the implementation of a solid waste treatment process provide a solution?

5 ADVANCED TREATMENT POTENTIAL

Currently bioremediation provides a large-scale, long-term option for relatively slow microbial degradation of a wide variety of compounds requiring low energy and low-level maintenance (as demonstrated in Fig. 2). Timescales, process security and public safety may prevent this option providing a solution of CD&E waste streams (particularly for hazardous wastes). Landfarming (a form of bioremediation which requires a higher level of engineering and maintenance than other forms of bioremediation) also shows promise but will require further adaptation to ensure public safety due to the tilling process to provide aeration. In terms of public health and safety, traditional landfills provide a high level of protection (in the short-term) due to their containment methods, however, their design accommodates low/no degradation of waste. In comparison bioreactor landfills have been designed to provide low level degradation and energy recovery potential, although these benefits come with management requirements. A major drawback to the use of bioremediation and landfarming, is the potential migration of very stable hazardous substances (e.g. asbestos), therefore containment to provide degradation is needed.

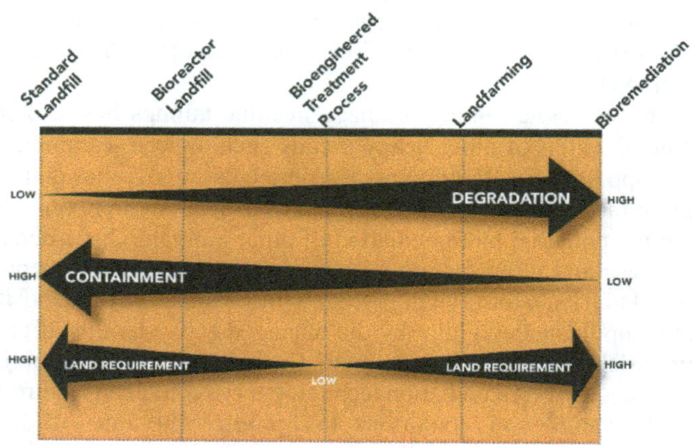

Figure 2: Disposal versus treatment in terms of microbial degradation, containment, land requirements.

In the centre of the two extremes (landfill versus bioremediation) lies the potential to create an engineered treatment process which could use aspects of current landfill design alongside optimised bioremediation (Bioengineered Treatment Process). This sustainable approach will require careful identification and separation of the waste types (Fig. 3), for which there must be a clear incentive. This process may require a pre-treatment stage and this should be followed by a well-managed waste process rather than a singular option of landfilling. Advantages of this type of process train may include:

- The creation of a less toxic environment due to the separation of hazardous substances could increase potential for microbial breakdown.
- A reduction in the creation of highly toxic hazardous mixtures which are difficult to manage in the event of an enclosure failure.
- The potential for reuse or recycling of hazardous substances (after processing), which completes the material loop, and is especially valuable for heavy metals.

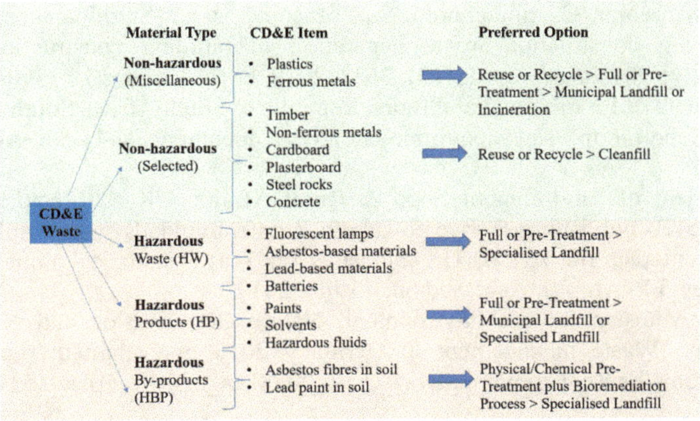

Figure 3: Defining and treating hazardous and non-hazardous CD&E waste.

6 CONCLUSIONS

We already have well-established biological treatment processes for the water and wastewater industry including suspended and fixed growth microbial systems. Although these systems have developed significantly over the last few decades, similar investments and advancements for solid waste treatment processes have not been forthcoming. This may be due to high land space requirements, lack of incentives and perceived lack of feasibility. However, the main reason is that landfill processes are simple and low cost by comparison despite their shortcomings in terms of sustainability.

The proposed bioengineered treatment process aims to expand and adapt the methods for organic waste bioreactor (or activated) landfills incorporating the treatment processes utilized by bioremediation, to a landfill design suitable for the treatment of even hazardous CD&E waste; to minimise risk to public but maximise treatment. We still need to be sure that short term risks are acceptable even though long-terms risks may be far better. For this purpose, knowledge about treatment options for organic and inorganic wastes must be combined in novel ways and assumptions tested. In the long-term, the development of multiple biotechnological processes (according to waste type) may provide a better and more sustainable solution to combined waste disposal in landfills.

ACKNOWLEDGEMENT

The authors would like to acknowledge the NZ Ministry for the Environment (MfE) Waste Minimisation Fund for part-funding this research.

REFERENCES

[1] Kaza, S., Yao, L., Bhada-Tata, P. & Van Woerden, F., *What a Waste 2.0: A Global Snapshot of Solid Waste Management to 2050*, World Bank: Washington, DC, 2018.

[2] United Nations Environmental Programme (UNEP), Sustainable building and construction: Facts and figures. *Industry and Environment – A Quarterly Review, United Nations Environmental Programme (April–September), Division of Technology, Industry and Economics*, **26**(2), pp. 5–98, 2003.

[3] Inglis, M., Construction and demolition waste: Best practice and cost saving. *Proceedings of the New Zealand Sustainable Building Conference*, 057, 2007.

[4] Luangcharoenrat, C., Intrachooto, S., Peansupap, V. & Sutthinarakorn, W., Factors influencing construction waste generation in building construction: Thailand's perspective. *Sustainability*, **11**(13), 3638, 2019. https://doi.org/10.3390/su11133638.

[5] Department of Environmental Affairs, Republic of South Africa, South Africa State of Waste Report. http://sawic.environment.gov.za/documents/8641.pdf. Accessed on: 13 Jun. 2019.

[6] Department of Environment Food & Rural Affairs, UK, UK statistics on waste. https://assets.publishing.service.gov.uk/government/uploads/system/uploads/attachment_data/file/784263/UK_Statistics_on_Waste_statistical_notice_March_2019_rev_FINAL.pdf. Accessed on: 13 Jun. 2019.

[7] Federal Ministry for the Environment, Nature Conservation and Nuclear Safety, Germany, Waste management in Germany 2018. www.bmu.de/fileadmin/Daten_BMU/Pools/Broschueren/abfallwirtschaft_2018_en_bf.pdf. Accessed on: 13 Jun. 2019.

[8] Blue Environment, National waste report 2018. Prepared for Australian Government, Department of the Environment and Energy. www.environment.gov.au/system/files/resources/7381c1de-31d0-429b-912c-91a6dbc83af7/files/national-waste-report-2018.pdf. Accessed on: 13 Jun. 2019.

[9] Eurostat, Waste statistics: Statistics explained. https://ec.europa.eu/eurostat/statistics-explained/pdfscache/1183.pdf. Accessed on: 30 Jun. 2019.

[10] Yonetani, H., Construction and demolition waste management in Japan. www.uncrd.or.jp/content/documents/2661Parallel%20Roundtable(2)-Presentation(4)-Hideko%20Yonetani.pdf. Accessed on: 30 Jun. 2019.

[11] Ministry for the Environment (MfE) New Zealand, Module 1: Hazardous waste guidelines: Identification and record-keeping. www.mfe.govt.nz/sites/default/files/hazardous-waste-module-1.pdf. Accessed on: 13 Feb. 2020.

[12] Boyle, C.A., Solid waste management in New Zealand. *Waste Management*, **20**, pp. 517–526, 2000.

[13] Ministry for the Environment (MfE) New Zealand, The New Zealand waste strategy: Reducing harm, improving efficiency. www.mfe.govt.nz/sites/default/files/wastestrategy.pdf. Accessed on. 12 Mar. 2020.

[14] PBC Today, Dealing with hazardous waste. www.pbctoday.co.uk/news/planning-construction-news/hazardous-waste-construction/53765/. Accessed on: 16 Mar. 2020.

[15] Hyder Consulting, Encycle Consulting & Sustainable Resource Solutions, Construction and demolition waste status report: Management of construction and demolition waste in Australia. www.environment.gov.au/system/files/resources/323e8f22-1a8a-4245-a09c-006644d3bd51/files/construction-waste.pdf. Accessed on: 9 Jul. 2019.

[16] Kumar, S., Chiemchaisri, C. & Mudhoo, A., Bioreactor landfill technology in municipal solid waste treatment: An overview. *Critical Reviews in Biotechnology*, **31**(1), pp. 77–97, 2011.

[17] Stats NZ, Measuring New Zealand's progress using a sustainable development approach, 2008: Topic 8: Waste. http://archive.stats.govt.nz/browse_for_stats/snapshots-of-nz/Measuring-NZ-progress-sustainable-dev-%20approach/sustainable-development/waste.aspx#. Accessed on: 18 Feb. 2020.

[18] Ministry for the Environment (MfE) New Zealand, Environment New Zealand 2007: Chapter 6: Waste. www.mfe.govt.nz/publications/environmental-reporting/environment-new-zealand-2007/chapter-6-waste. Accessed on: 18 Feb. 2020.

[19] Ministry for the Environment (MfE) New Zealand, Solid waste composition: Environmental report card July 2009. www.mfe.govt.nz/environmental-reporting/waste/waste-composition-2009/index.html. Accessed on: 2 Mar. 2020.

[20] Ministry for the Environment (MfE) New Zealand, Types of landfills. www.mfe.govt.nz/waste/waste-guidance-and-technical-information/types-of-landfills. Accessed on: 18 Feb. 2020.

[21] Ministry for the Environment (MfE) New Zealand, Module 2: Hazardous waste guidelines – Landfill waste acceptance criteria and landfill classification. www.mfe.govt.nz/sites/default/files/haz-waste-guide-module-2-may04.pdf. Accessed on: 18 Feb. 2020.

[22] Auckland Council, Auckland's waste assessment 2017. www.aucklandcouncil.govt.nz/plans-projects-policies-reports-bylaws/our-plans-strategies/topic-based-plans-strategies/environmental-plans-strategies/docswastemanagementplan/waste-assessment-2017.pdf. Accessed on: 30 Jun. 2019.

[23] Ministry for the Environment (MfE) New Zealand, Targets in the New Zealand waste strategy: 2006 review of progress. www.mfe.govt.nz/publications/waste/waste-strategy-review-progress-mar07/html/page5a.html#figure14. Accessed on: 30 Jun. 2019.

[24] Carson, D.A., The municipal solid waste landfill operated as a bioreactor. *Proceedings of the Landfill Bioreactor Design and Operation Conference*, pp. 1–8, 1995.

[25] Centre for Advanced Engineering (CAE), Landfill guidelines: Towards sustainable waste management in New Zealand. www.mfe.govt.nz/sites/default/files/4139_landfill.pdf. Accessed on: 26 Feb. 2020.

[26] Waste Management Institute New Zealand (WasteMINZ), Technical guidelines for disposal to land. www.wasteminz.org.nz/wp-content/uploads/2016/04/Technical-Guidelines-for-Disposal-to-Land-9Aug18-FINAL.pdf. Accessed on: 2 Mar. 2020.

[27] Zero Waste New Zealand, Waste opportunity: A closer look at landfilling and incineration. http://zerowaste.co.nz/assets/Wastedopportunities.pdf. Accessed: 5 Mar. 2020.

[28] Tohill, S., Personal communication, February 2020. New Zealand Demolition and Asbestos Association (NZDAA), New Zealand.

[29] Thom, N.G., Factors influencing the satisfactory disposal of special wastes. *Transactions of the Institute of Professional Engineers New Zealand: Civil Engineering Section*, **12**(1), pp. 38–46, 1985.

[30] Miller, N., What can you do with a problem landfill? *World Wastes*, **39**(2), p. 42, 1996.

[31] Merry, S.M., Kavanzanjian Jr., E. & Fritz, W.U., Reconnaissance of the July 10, 2000, Payatas landfill failure. *Journal of Performance of Constructed Facilities*, **19**(2), pp. 100–107, 2005.

[32] Lavigne, F. et al., The 21 February 2005 catastrophic waste avalanche at Leuwigajah dumpsite, Bandung, Indonesia. *Geoenvironmental Disasters*, **1**(10), 2014. https://doi.org/10.1186/s40677-014-0010-5.

[33] BBC News, Brazil landslide death toll rises. http://news.bbc.co.uk/2/hi/8612699.stm. Accessed on: 5 Mar. 2020.

[34] Petley, D., Garbage dump landslide in the Philippines. https://blogs.agu.org/landslideblog/2011/08/31/garbage-dump-landslide-in-the-philippines/. Accessed on: 5 Mar. 2020.

[35] Musulin, K., Collapse at Guatemala City's largest garbage dump leaves 4 dead, 24 missing. www.wastedive.com/news/collapse-at-guatemala-citys-largest-garbage-dump-leaves-4-dead-24-missing/418454/. Accessed on: 5 Mar. 2020.

[36] Petley, D., Koshe, Ethiopia: The worst garbage dump landslide in recent years. https://blogs.agu.org/landslideblog/2017/03/17/koshe-1/. Accessed on: 5 Mar. 2020.
[37] Daily Mirror, Meethotamulla tragedy. www.dailymirror.lk/article/Meethotamulla-tragedy--127303.html. Accessed on: 5 Mar. 2020.
[38] Sesay, I., Search for two people buried alive under landslide suspended due to asbestos fear. www.euroweeklynews.com/2020/02/07/breaking-update-search-for-two-people-buried-alive-under-landslide-suspended-due-to-asbestos-fear/. Accessed on: 3 Mar. 2020.
[39] Ewing, I., Environmental catastrophe after West Coast storm exposes old landfill. www.newshub.co.nz/home/new-zealand/2019/04/environmental-catastrophe-after-west-coast-storm-exposes-old-landfill.html. Accessed on: 14 Feb. 2020.
[40] NZ Herald, Fears sea will breach old landfill in Dunedin, creating rubbish disaster. www.nzherald.co.nz/index.cfm?objectid=12256976&ref=twitter. Accessed on: 13 Feb. 2020.
[41] Carrington, D., Pollution risk from over 1,000 old UK landfill sites due to coastal erosion. www.theguardian.com/environment/2016/may/05/pollution-risk-from-over-1000-landfill-sites-england-wales-coastal-erosion. Accessed on: 26 Feb. 2020.
[42] Wharton, J., Landfill site starts leaking plastic into the Thames. https://metro.co.uk/2019/11/06/landfill-site-starts-leaking-plastic-thames-11050538/. Accessed on: 26 Feb. 2020.
[43] 7News, Residents living above Melbourne's Sunshine Landfills potentially leaking toxic gas. https://7news.com.au/news/environment/residents-living-above-melbournes-sunshine-landfills-potentially-leaking-toxic-gas-c-720854. Accessed on: 5 Mar. 2020.
[44] Davis, M.L. & Cornwell, D.A., *Introduction to Environmental Engineering*, 5th ed., McGraw-Hill: New York, 2012.
[45] Spasiano, D., Dark fermentation process as pretreatment for a sustainable denaturation of asbestos containing wastes. *Journal of Hazardous Materials*, **349**, pp. 45–50, 2018. https://doi.org/10.1016/j.jhazmat.2018.01.049.
[46] Vidali, M., Bioremediation: An overview. *Pure Applied Chemistry*, **73**(7), pp. 1163–1172, 2001.
[47] Nishiwaki, J., Kawabe, Y., Komai, T. & Zhang, M., Decomposition of gasoline hydrocarbons by natural microorganisms in Japanese soils. *Geosciences*, **8**(2), p. 35, 2018. https://doi.org/10.3390/geosciences8020035.
[48] Tolaymat, T., Kim, H., Jain, P., Powell, J. & Townsend, T., Moisture addition requirements for bioreactor landfills. *Journal of Hazardous, Toxic, and Radioactive Waste*, **17**(4), pp. 360–364, 2013.
[49] Arafat, H.A., Jijakli, K. & Amimul, A., Environmental performance and energy recovery potential of five processes for municipal solid waste treatment. *Journal of Cleaner Production*, **105**, pp. 233–240, 2015.
[50] Blackburn, J.W., Bioremediation scale-up effectiveness: A review. *Bioremediation*, **1**(4), pp. 265–282, 1998.
[51] Das, S. & Dash, H.R., Microbial bioremediation: A potential tool for restoration of contaminated areas. *Microbial Biodegradation and Bioremediation*, ed. S. Das, Elsevier Insights: USA, pp. 1–21, 2014.
[52] Karigar, C.S. & Rao, S.S., Role of microbial enzymes in the bioremediation of pollutants: A review. *Enzyme Research*, 2011. https://doi.org/10.4061/2011/805187.

[53] Sanchez-Castro, I. et al., Screening of bacterial strains isolated from uranium mill tailing porewaters for bioremediation purposes. *Journal of Environmental Radioactivity*, **166**, pp. 130–141, 2017. https://doi.org/10.1016/j.jenvrad.2016.03.016.

[54] Merroun, M.L. & Selenska-Pobell, S., Bacterial interactions with uranium: An environmental perspective. *Journal of Contaminant Hydrology*, **102**(3/4), pp. 285–295, 2008. https://doi.org/10.1016/j.jconhyd.2008.09.019.

[55] Karaca, O., Camaselle, C. & Reddy, K.R., Mine tailing disposal sites: Contamination problems, remedial options and phytocaps for sustainable remediation. *Reviews in Environmental Science and Biotechnology*, **17**(1), pp. 205–228, 2017. https://doi.org/10.1007/s11157-017-9453-y.

[56] Wallis, S.L. et al., Challenging global waste management: Bioremediation to detoxify asbestos. *Frontiers in Environmental Science*, **8**(20), pp. 1–14, 2020. https://doi.org/10.3389/fenvs.2020.00020.

[57] Federal Remediation Technologies Roundtable (FRDF), Remediation technologies screening matrix and reference guide, Version 4.0, 4.13 Landfarming. https://frtr.gov/matrix2/section4/4_13a.html. Accessed on: 20 Nov. 2019.

[58] Center for Public Environmental Oversight (CPEO), Landfarming. www.cpeo.org/techtree/ttdescript/lanfarm.htm. Accessed on: 20 Nov. 2019.

[59] Daghino, S., Martino, E., Fengolio, I., Tomatis, M., Perotto, S. & Fubini, B., Inorganic materials and living organisms: Surface modifications and fungal responses to various asbestos forms. *Chemistry*, **11**(19), pp. 5611–5618, 2005. https://doi.org/10.1002/chem.200500046.

[60] Daghino, S. et al., Bioweathering of chrysotile by fungi isolated in ophiolitic sites. *FEMS Microbiology Letters*, **285**(2), pp. 242–249, 2008. https://doi.org/10.1111/j.1574-6968.2008.01239.x.

DEVELOPMENT OF AN ANALYTICAL REACTION KINETICS MATHEMATICAL MODEL BASED ON THERMOGRAVIMETRIC DATA FOR RECLAIMED PLASTIC WASTE FROM ACTIVE LANDFILLS

SULTAN MAJED AL-SALEM[1], HAJAR JAWAD KARAM[1], MAJED HAMEED AL-WADI[2] & GARY ANTHONY LEEKE[2]
[1]Environment and Life Sciences Research Centre, Kuwait Institute for Scientific Research (KISR); Kuwait
[2]School of Chemical Engineering, University of Birmingham, UK

ABSTRACT

Landfill mining (i.e. waste reclamation) has been receiving renewed attention in research circles. This strategy provides a solution to the overgrowing problem of land use for the classical solution of landfilling. In addition, it provides added value for the materials recovered with a potential of boosting local economies, especially considering *circular economy* market development. In this work, the degradation kinetics were investigated for solid waste retrieved from three active landfill sites. The landfill sites operate unsanitary processing schemes for the management of the municipal solid waste they receive. Thermogravimetric analysis was used to study the micro-kinetics using standardized ICTAC methods, which yielded an apparent activation energy (E_a) for one of the sites in the range of 100 to 121 kJ mol^{-1} using the Criado method. Another site showed a range between 87 to 107 kJ mol^{-1} using the same method. The Avrami-Erofeev expression for degradation mechanism was identified as the prevailing one based on the analysis conducted. Furthermore, an analytical solution model based on reaction order and degradation curves' initial and final set-points was developed to yield the desired parameters in a more realistic manner. We conclude that the type of waste, period of burial and sensitivity of kinetic models used are the most prevailing parameters in determining reaction mechanisms and kinetics expressions, which effect the development of industrial units that use thermal kinetics data for their design of treating various waste components.
Keywords: landfilling, thermogravimetry, landfill mining, kinetics, thermolysis.

1 INTRODUCTION

Unsanitary landfilling that utilises open dump sites, is still practiced the world over as a waste management method to dispose of municipal solid waste (MSW). This practice is still undertaken in many developing world countries namely on the Asian continent [1]–[3]. This practice is also associated with various adverse environmental effects such as leachate infiltration to groundwater [4]–[5] and landfill occupation beyond operational lifespan [6]. Landfilling plastic solid waste (PSW) is also considered to be one of the main environmental threats that could compromises a site for future rehabilitation and causes various economical loss due to lack of proper management [7]–[11]. The majority of these plastics are thermoplastic in type and could be valorised in thermo-chemical treatment (TCT) units to produce various types of fuels and energy [12]–[14]. Prime examples of such plastics are low density (LDPE) and high-density polyethylene (HDPE), polypropylene (PP), and polyvinyl chloride and alcohol (PVC and PVA). Readers are referred to Al-Salem et al. [15]–[17] for a review on the state of art regarding TCT technologies. One of the main solutions that present itself as an advantageous option for land rehabilitation and reduction of environmental burdens, is landfill mining [18]–[21]. The reclamation of the MSW from the landfill site would reduce the occupied space and could potentially be used as a feedstock material for various fuel and energy production processes [22], [23]. Various research efforts the world over are reporting on PSW

WIT Transactions on Ecology and the Environment, Vol 247,
www.witpress.com, ISSN 1743-3541 (on-line)
doi:10.2495/WM200171

reclamation processes and their potential use post mining of landfill sites [23], [24]. On the other hand, technical literature is quite lacking a major element that can determine the appropriateness of TCT application of such waste. The fact that basic material degradation kinetics are lacking for PSW reclaimed from landfill sites, is a major research gap that needs to be addressed.

In this work, thermogravimetric analysis (TGA) was used to study the micro-kinetics of PSW reclaimed from two active landfill sites by applying the standardized international confederation for thermal analysis and calorimetry (ICTAC) methods [25], [26]. The methods of Criado and the method of Al-Salem and Khan [27] which is based on an analytical degradation reaction kinetics model used for solving complex reaction systems, were used in this work. The results are considered to be the cornerstone for reactor design and simulation of such results for industrial scale-up in the future.

2 MATERIALS AND METHODS

The waste was mined from two active unsanitary landfill site, namely Mina Abdullah (MAB) (lat. 29° 19' 33.24" N; long. 47° 36' 41.04" E) and Al-Jahra (JAH) (lat. 28° 59' 54.564" N; long. 48° 6' 25.146" E) landfill sites within boarders of the State of Kuwait. Readers are referred to Al-Salem et al. [24] for a detailed map location of the sites and to Al-Salem et al. [28] and Al-Jarallah and Aleisa [29] for more information on the daily activities and operation of the sites. The mining and reclamation of the buried SW was conducted following internationally recognised protocols previously published and applied in research and development works [11], [24], [30]. The SW was mined from a singular ditch where the MSW was buried for six months (under 1 meter of soil), and the sampling of the material was conducted after removing the topsoil (≈10 cm) [31]. Details on the landfill site and pollutants levels including soil characteristics are shown elsewhere [32]–[34]. Fig. 1 shows a pictorial depiction of the process giving an overall view and black matter was noticed of the reclamation/mining process. The reclaimed waste (30 kg) was air blown, washed and conditioned in accordance with previous waste mining protocols [24], [31]. All samples during this work were stored in laboratory conditions (22–23°C/50% relative humidity) and kept in sealed containers. The PSW was shredded using a Vema Company machine operated at 580 rpm speed (mesh size: 5–15 mm) [31]. The majority of the reclaimed waste was PSW (> 20%) for both sites as reported in Al-Salem [35]. The materials were then subjected to cryogenic milling to a particle size using the protocol shown previously in Al-Salem et al. [24] to give particles of c. 50 mm in diameter which was verified using micrograph analysis [24], [31]. Table 1 shows an example of the mass of samples reclaimed (JAH site), and the percentile analysis of all waste categories in this work is shown in Table 2 for JAH site as an example. Elemental Analysis was conducted by dynamic flash combustion using a 2±0.1 mg in a Thermo Flash 2000 analyser as per ASTM D5373 [36] and ASTM D5291 [37]. Ash Content was determined using gravimetric analysis with a muffle furnace as per ASTM D5630 [38] for plastics and ISO/DIS 1762 [39] for the adjusted temperature of 525±25°C by testing 4 ± 0.1 g of the reclaimed PSW (Table 3).

Thermal degradation of the samples was investigated using a Mettler-Toledo TGA 3+ Model set to record the data every second under multiple heating rates (β) (i.e. 5, 10, 15, 20 and 25°C min^{-1}) conforming with the approved methods of the ICTAC [25], [26]. These were conducted to diminish sample size influences on the weight loss and kinetic analysis. A constant flow of nitrogen (N_2, dry–99.99% purity) with a flow rate of 50 ml min^{-1} was maintained throughout the experiments. The measurements were conducted in triplicate using 1±0.1 mg samples from room temperature (RT) to 600°C and showed high repeatability. The crucibles used in all experimental runs were made of alumina with a 70 µl size. Mass loss (%) and first derivative (DTG, % °C^{-1}) were recorded constantly to the properties of the thermogram. The TGA unit was externally calibrated and maintained regularly before each experimental batch using standard reference materials as per ISO 11358 [40].

Figure 1: Pictorial depiction of the landfill mining process showing (left) soil top view of JAH landfill site, (middle) ditch excavation process at MAB, and (right) spade tractor operation/excavation of waste at MAB.

Table 1: Weight measured (kg) during the shredding of waste samples (Al-Jahra landfill site).

Date of sampling	Site		Flake size	Shredding cycles
4 Nov. 2018	JAH landfill site		5–90 mm	3
	Initial weight	After washing	After drying	After shredding
Plastic	6	12.48	3.7	3.45
Metal	1.4	1.4	1.2	
Paper	1.4	/	1.2	
Wood	8.2	/	8.2	
Organic	0	/	0	
Others	5.1	/	5	

Table 2: Percentile analysis of waste reclaimed from Al-Jahra landfill site.

Waste type	Percentile before washing	Waste type	Percentile after drying
Plastic	27.1	Plastic	19.1
Metal	6.3	Metal	6.2
Paper	6.3	Paper	6.2
Wood	37.1	Wood	42.4
Organic	0.0	Organic	0.0
Others	23.0	Others	25.9

Table 3: Elemental analysis and physio-chemical properties conducted on reclaimed plastic solid waste.

Sample	Total nitrogen (TN) %	Total carbon (TC) %	Total hydrogen (TH) %	Moisture at 105°C
Al-Jahra	0.123 ± 0.02	75.137 ± 0.09	11.412 ± 0.35	0.155 ± 0.02
MAB	0.084 ± 0.00	77.902 ± 0.16	12.968 ± 0.15	0.005 ± 0.00
Sample	Moisture at 105°C	Organic matter at 550°C	Ash at 550°C	Oxygen (O) %
Al-Jahra	0.155 ± 0.02	93.795 ± 0.12	6.050 ± 0.10	8.85± 0.00
MAB	0.005 ± 0.00	95.180 ± 0.06	4.815 ± 0.06	6.27± 0.60

3 DEGRADATION REACTION KINETICS ANALYSIS

The plastic fraction under weight loss investigation is reported with respect to the conversion of the polymeric material (α) thus [41]–[43]

$$\alpha = \frac{m - m_o}{m_o - m_f}, \tag{1}$$

where m, m_o and m_f are the mass of polymeric material at a specific reaction time (t), initial and final mass of the polymer at the investigated temperature, respectively. The degradation rate was defined with respect to rate of reaction by incorporating a first order Arrhenius equation [27], [43]–[46]

$$d\alpha/dt = A_o \exp\left(\frac{-E_a}{RT}\right) f(\alpha), \tag{2}$$

where A_o stands for the frequency factor (min^{-1}), E_a is the apparent activation energy (kJ mol^{-1}), T is the reaction temperature at desired time (K) and $f(\alpha)$ is the reaction model that represents the degradation mechanism following reaction order (n). Multiple heating rate (β) values were used in this work as recommended by various authors for accuracy reasons [44]. ICTAC Kinetics Committee recommendations for the application of Criado's approach has identified that the Avrami-Erofeev expression for $f(\alpha)$ be applied for the estimation of Arrhenius parameters for polymer degradation kinetics from the following:

$$\frac{d\alpha}{dt.f(\alpha)} = f(T),\ \int \frac{d\alpha}{\left(2(1-\alpha).\sqrt{-\ln(1-\alpha)}\right)} = \int A_o e^{-\frac{E}{RT}}.dt. \tag{3}$$

Apparent activation energy calculated by this method showed similar trends. The reader is referred to Al-Salem and Khan [27] for the full derivation of the model. The LHS of the eqn was plotted against the inverse of Temperature in K and the slope yielded the energy of activation and pre-exponential factor for each set of experiments. The whole exercise was repeated for the estimation of non-isothermal kinetics of all samples where n is not fixed but optimized

$$\frac{d\alpha}{dt} = -A_0 e^{-\left(\frac{E}{RT}\right)}.\alpha_n. \tag{4}$$

Eqn (4) is the rate expression for non-isothermal, if n =1

$$\alpha = \exp\left(-A_0.e^{-\left(\frac{E}{RT}\right)}.t\right) \text{ and for } n \neq 1,\ \alpha = \left((n-1).A_0.e^{-\left(\frac{E}{RT}\right)}.t+1\right)^{-(1/(n-1))}. \tag{5}$$

The experimental data were smoothed using mathematical function of first order kinetics presented previously and was normalized using initial and final weights of the experimental dataset (W_{t0} and W_{tfinal}) for 5°C/min as seen below. The reader is referred to Al-Salem and Khan [27] for the full derivation of the analytical solution model

$$W(t) = Exp\left(-A_0 . \exp\left(-\frac{E}{RT}\right) . t\right)\left(W_{t0} - W_{tfinal}\right) + W_{tfinal}. \tag{6}$$

The smoothed data plots for three landfill sites presented in the next section were used for testing different reaction kinetics models. The exact reaction mechanism based on the best fit to the smoothed experimental data was selected and all the chemical kinetic parameters were evaluated.

4 RESULTS AND DISCUSSION

Thermal characterization of the reclaimed PSW was conducted using TGA describing the thermal decomposition of the material by measuring the weight loss as function of sample temperature or reaction time under isothermal conditions. It also determines the temperatures that depicts the steps (weight loss/change). The thermal behaviour of the material identified from the TGA thermogram and DTG (i.e. first derivative curve) are essential for evaluating the thermal stability of the material from which the thermal profile is also identified. The experimental campaign included samples of PSW from different landfills. The comparison with previous results from literature was made. Samples in this work have also been tested for pyrolytic properties and estimation of the kinetic parameters under five heating ramps. The resulting data helps in determining the influence of the effect of burying of plastics on the physio-chemical properties and thermally degradation characteristics. For further discussion on the aforementioned properties, the reader is referred to the cited references of our previous work [24], [31], [35].

TGA data of used plastic from three landfill sites were studied using smoothening techniques and were used for the evaluation of decomposition reaction parameters. Fig. 2 shows the raw experimental TGA data for PSW tested from the MAB and JAH sites. The result of MAB site shows that a one-step mass loss exists in the beginning due to volatilization in the temperature range from 230 to 350°C. A minor second degradation step was also detected within the temperature range of 370 to 450°C. JAH site samples showed a mass loss due to volatilization in the temperature range from 250 to 360°C (Fig. 2). In the work of Tuffi et al. [47] the thermal analysis of polymeric mixtures from waste packaging plastics was performed. The mixture mainly consists of waste polymers (PP, polyethylene film (PE), PET and PS). In addition, the thermal behaviour of three synthetic mixtures demonstrating commingled postconsumer plastics wastes (CPCPWs) from material recovery facilities, were considered. A 10 mg of powder samples in alumina crucibles was prepared and tested in their work. The thermal analysis experiments were carried out under a N_2 flow rate of 60 ml/min in the temperature range 25 to 600°C. The degradation temperatures of single polymers and their mixtures were determined at heating rate 10°C/min. The elemental and proximate analysis of the samples were performed and gave volatile matter of approximately 100% except for PET with a very low ash content is. The TG curves of the four polymers show that the degradation occurred in the temperature range 400 to 500°C) in a one-step mass loss. Furthermore, the onset temperatures for PE film, PP, PS and PET were (which are equivalent to 477, 456, 411, 434°C respectively). This is in-line with TGA analysis conducted in this work implying that the samples are of Polyolefin PE origin as supported by the thermal profile depicted in Table 4. The thermal behaviour of the synthetic mixtures, CPCW1, CPCW2 and CPCW3 were also determined.

The degradation occurred in two steps mass losses for CPCW2 and CPCW3 where in CPCW1 is one-step. The onset temperature for CPCW1 was 452.85°C. The first step of CPCW2 and CPCW3 in a similar range to the degradation temperature for PS and PET (418°C) where, the second step is close to PP and PE film (461–465°C). The experiments occurred at different heating rates (2, 5, 10 and 15°C/min).

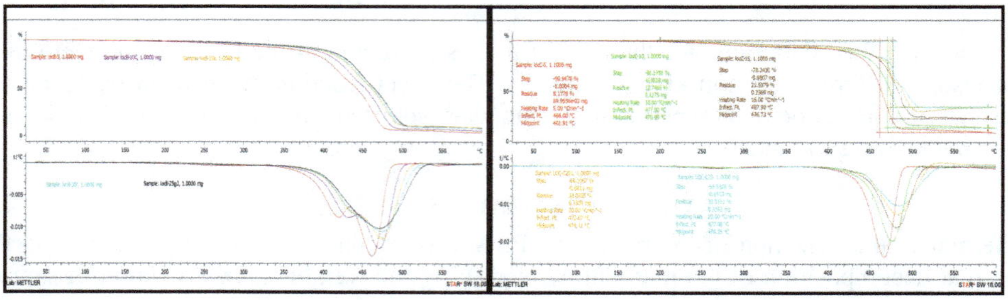

Figure 2: Experimental TGA results for MAB (left) and JAH (Right) landfill sites showing the thermogram and first derivative (DTG) curves.

Table 4: Thermal degradation and stability profile showing onset temperature (Tos, °C), inflection point (T_{if}, °C), midset temperature (T_{ms}, °C), and end of set temperature (T_{ed}, °C).

β (°C/min)	T_{os} °C	T_{if} °C	T_{ms} °C	T_{ed} °C	T_{os} °C	T_{if} °C	T_{ms} °C	T_{ed} °C
5	403.9	454.1	448.1	498.6	404.5	453.5	449.3	498.0
10	415.0	466.0	460.83	513.1	415.8	466.3	462.0	512.5
15	421.7	474.5	468.5	522.0	422.5	473.5	469.7	521.2
20	426.6	482.0	474.0	528.3	427.3	480.0	475.3	527.6
25	430.41	484.1	478.33	533.3	431.2	484.5	479.5	532.5

Blazevska-Gilev and Spaseska [48] studied the thermal degradation of polyvinyl chloride by dynamic thermogravimetry (TGA), up to around 450°C at heating rates between 4 and 10°C/min using 100 ml/min of argon gas. The first degradation step ended at around 300°C, with a weight loss of about 65%, while the second step ended with a 85% total weight loss. It has been reported that the degradation of PVC is a more complex than the other plastic [48]. Thermal analysis for the degradation of starch/poly (vinyl alcohol) (PVA) blends were carried out using TGA under atmospheric conditions by Taghi Taghizadeh and Abdollahi [49]. Their dynamic experiments were conducted at different heating rates (5, 10 and 15°C/min) from RT to 600°C. The pure starch and starch/PVA blend curves showed a three-step decomposition curve. The first step mass loss was around 260°C due to the loss of volatile matter and water. The second step mass loss occurred in the temperature range of 260°C to 360°C described the polymer decomposition and was attributed to deterioration of the polymer's backbone. Another study in our earlier work investigated the TGA degradation of PVA [50]. The experiment was conducted between 40 to 700°C at a heating rate of 10°C/min under N_2 flow rate of 50 ml/min. The first step mass loss was due to the water content in the sample at temperature of 100°C. At a temperature around 300°C, the sample started to further degrade because of the acetate group with a third step between 350°C and 450°C, which related to the PVA degradation rate.

Furthermore, the onset temperature was evaluated to be 301±2°C with a T_{peak} at 334 to 440°C [50]. Kuźnia and Magdziarz [51] studied three types of PSW based polyolefins as well using a combustion set-up. The beginning of the degradation process was around 250°C due to the combustion of the alkane groups; the temperature of spontaneous combustion is lower than 250°C for alkanes such as n-hexane and n-heptane, n-octane, n-nonane, n-decane and n-undecane. The end-set temperature of PE was estimated to be 550°C and approximately 400°C for PP [51]. Tables 5–8 and Fig. 3 show some of the results from the mathematical modelling exercise executed in this work. The average value of Activation Energy for the MAB samples based on n^{th} order kinetics model was 260 kJ mol^{-1} and 240 kJ mol^{-1} based on the 1st order model, where the optimum value of n was equal to 1.1. The average value of Activation Energy using the Criado model for the MAB sample is 109.828 kJ mol^{-1}, while the average activation energy for the JAH sample is 240 kJ mol^{-1} using the first order reaction model. The average Activation Energy of the JAH samples based on n^{th} order kinetics model was 230 kJ mol^{-1} where the optimum value of n was equal to 1.1. The average value of Activation Energy using Criado model for JAH is 97.922 kJ mol^{-1}. MAB samples showed a single thermal degradation step that could be presented by a general pyrolytic reaction. The JAH samples clearly followed two stages of thermal degradation showing similar behaviour as past reports [52]. The data indicate that there are different mechanisms of degradation for plastics buried in sand in dry and arid climates. Al-Salem and Khan [27] have analysed the TGA data of polymeric blends of different plastics in various proportions. They reported two sets of Arrhenius coefficients of parallel reactions. Agrawal [53] has proposed simple reaction mechanisms for non-isothermal kinetics of thermal degradation which are strongly influenced by physical factors such as particle size, and geometry of the sample holder. These physical factors could be defined as diffusion-limited or by other mechanisms that strongly influence $f(\alpha)$, which are crucial for the evaluation of kinetic parameters for the reaction model. Zong et al. [52] have reported thermal degradation kinetics of polyethylene and silane-crosslinked polyethylene samples based on n^{th} order of reaction. The orders of reaction were lower than reported in the literature, ranging from 0.24 to 0.59 and the corresponding Activation Energy values were similar to those reported in the literature ranging between 226 to 294 kJ mol^{-1}. Further results extracted from similar modelling approach on polyolefin polymers are depicted elsewhere [41], [42].

Table 5: Summary of obtained kinetic parameters for MAB site using present mathematical model.

Ramp	Pre-exponential (s^{-1})	Activation energy (kJ mol^{-1})
5°C/min	2.17678×10^{16}	280
10°C/min	9.74489×10^{15}	275
15°C/min	4.42434×10^{15}	270
20°C/min	2.02573×10^{15}	265
25°C/min	9.32373×10^{14}	260

Table 6: Summary of obtained kinetic parameters for MAB site using present mathematical model with reaction order equal to 1.1 and Criado method results.

Ramp for n = 1.1 expression	Pre-exponential ($\alpha^{-0.1}s^{-1}$)	Activation energy (kJ mol^{-1})	Activation energy from Criado method kJ mol^{-1}
5°C/min	$4.93x10^{15}$	270	121.46
10°C/min	$2.29x10^{15}$	265	119.80
15°C/min	$1.09x10^{15}$	260	104.25
20°C/min	$5.08x10^{14}$	255	102.76
25°C/min	$2.41x10^{14}$	250	100.84

Table 7: Summary of obtained kinetic parameters for JAH site using present mathematical model.

Ramp	Pre-exponential (s^{-1})	Activation Energy (kJ mol^{-1})
5°C/min	$8.82x10^{13}$	250
10°C/min	$4.00x10^{13}$	245
15°C/min	$1.84x10^{13}$	240
20°C/min	$8.57x10^{12}$	235
25°C/min	$4.01x10^{12}$	230

Table 8: Summary of obtained kinetic parameters for JAH site using present mathematical model with reaction order equal to 1.1 and Criado method results.

Ramp for n = 1.1 expression	Pre-exponential ($\alpha^{-0.1}s^{-1}$)	Activation energy (kJ mol^{-1})	Activation energy from Criado method kJ mol^{-1}
5°C/min	$1.99x10^{13}$	240 kJ mol^{-1}	107.08
10°C/min	$9.04x10^{12}$	235 kJ mol^{-1}	106.50 kJ mol^{-1}
15°C/min	$4.48x10^{12}$	230 kJ mol^{-1}	99.60 kJ mol^{-1}
20°C/min	$2.16x10^{12}$	225 kJ mol^{-1}	89.04 kJ mol^{-1}
25°C/min	$1.03x10^{12}$	220 kJ mol^{-1}	87.38 kJ mol^{-1}

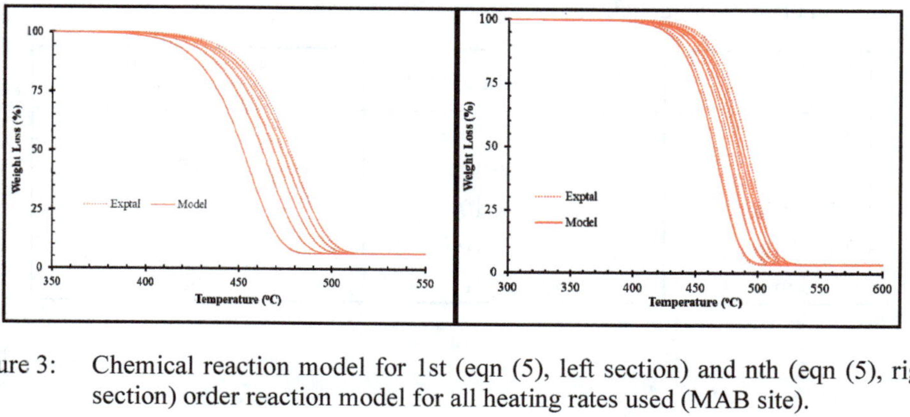

Figure 3: Chemical reaction model for 1st (eqn (5), left section) and nth (eqn (5), right section) order reaction model for all heating rates used (MAB site).

5 CONCLUSION

Two landfill sites that are characterised in being both unsanitary and in located in an arid-land, were mined and the solid waste from both sites was reclaimed. The plastic waste was conditioned and subjected to thermogravimetry to study its thermal stability and behaviour. The thermal properties detected were in-line with past efforts; and showed that the materials were of polyolefin polymer origin namely polyethylene. The degradation was modelled by two kinetic of reaction models were, using an analytical solution model that smoothened the thermograms and determined the properties based on optimising the weight loss steps; and a second solution based on the Criado approach. The analytical model showed more realistic results following the mass losses detected and based on the activation energy estimated. This is deduced based on the similarity of energies estimated and the estimation of the model for each thermal loss (step) on its own, accounting for each polymer's degradation as a single loss. Furthermore, the results of both kinetic approaches shows that the materials have similar properties to virgin polyolefin polymers, making it appropriate to further design and investigate the scale-up opportunities of the thermolysis design for mined waste to fuel and energy. This also paves the way for future plans in managing sold waste from various landfill sites to reduce the burdens on land and resources.

ACKNOWLEDGEMENTS

The authors would like to thank the Kuwait Institute for Scientific Research (KISR) and the British Council/UK-Gulf Institutional Links for funding research project EM103C. We would also like to thank Kuwait Municipality (KM) for their support towards field work activities reported in this communication. The authors would like to thank the Kuwait Foundation for the Advancement of Sciences (KFAS) for funding this scientific mission.

REFERENCES

[1] Al-Salem, S.M., Abraham, G., Al-Qabandi, O.A. & Dashti, A.M., Investigating the effect of accelerated weathering on the mechanical and physical properties of high content plastic solid waste (PSW) blends with virgin linear low density polyethylene (LLDPE). *Polymer Testing*, **46**, pp. 116–121, 2015.

[2] Singh, A. & Chandel, M.K., Effect of ageing on waste characteristics excavated from an Indian dumpsite and its potential valorisation. *Process Safety and Environmental Protection*, **134**, pp. 24–35, 2020.

[3] Kumar, S. et al., Challenges and opportunities associated with waste management in India. *Royal Society Open Science*, **4**, 160764, 2017.

[4] Pujari, P.R. & Deshpande, V., Source apportionment of groundwater pollution around landfill site in Nagpur. *Indian Environmental Monitoring & Assessment*, **111**, pp. 43–54, 2005.

[5] Mor, S., Ravindra, K., Dahiya, R.P. & Chandra, A., Leachate characterization and assessment of groundwater pollution near municipal solid waste landfill site. *Environmental Monitoring & Assessment*, **118**, pp. 435–456, 2006.

[6] Sharholy M., Ahmad K., Mahmood G. & Trivedi R.C., Municipal solid waste management in Indian cities – A review. *Waste Management*, **28**, pp. 459–467, 2008.

[7] Canopoli, L., Coulon, F. & Wagland, S.T., Degradation of excavated polyethylene and polypropylene waste from landfill, *Science of the Total Environment*, **698**, pp. 134125, 2020.

[8] Canopoli, L., Fidalgo, B., Coulon, F. & Wagland, S.T., Physico-chemical properties of excavated plastic from landfill mining and current recycling routes. *Waste Management*, **76**, pp. 55–67, 2018.

[9] He, P., Chen, L., Shao, L., Zhang, H. & Lü, F., Municipal solid waste (MSW) landfill: A source of microplastics? Evidence of microplastics in landfill leachate. *Water Research*, **159**, pp. 38–45, 2019.

[10] Kamura, K., Omori, M., Kurokawa, M. & Tanaka, H., Estimation of the potential of landfill mining and exploration of metal-enriched zones. *Waste Management*, **93**, pp. 122–129, 2019.

[11] Quaghebeur, M. et al., Characterization of landfilled materials: screening of the enhanced landfill mining potential. *Journal of Cleaner Production*, **55**, pp. 72–83, 2013.

[12] Al-Salem, S.M., Thermal pyrolysis of high density polyethylene (HDPE) in a novel fixed bed reactor system for the production of high value gasoline range hydrocarbons (HC). *Process Safety & Environmental Protection*, **127**, pp. 171–179, 2019.

[13] Al-Salem, S.M., Introduction, Chapter 1. *Plastics to Energy: Fuel, Chemicals & Sustainable Implications*, ed S.M. Al-Salem, Elsevier. https://doi.org/10.1016/B978-0-12-813140-4.00001-7, 2019.

[14] Al-Salem, S.M., Kinetic studies related to polymer degradation and stability. *Plastics to Energy: Fuel, Chemicals, and Sustainability Implications*, Elsevier, 2019, pp. 233–268. https://doi.org/10.1016/B978-0-12-813140-4.00009-1, 2019.

[15] Al-Salem, S.M., Antelava, A., Constantinou, A., Manos, G. & Dutta, A., A review on thermal and catalytic pyrolysis of plastic solid waste (PSW). *Journal of Environmental Management*, **197**, pp. 177–198, 2017.

[16] Al-Salem, S.M., Lettieri, P. & Baeyens, J., Recycling and recovery routes of plastic solid waste (PSW): A review. *Waste Management*, **29**(10), pp. 2625–2643, 2009.

[17] Al-Salem, S.M. et al., A review of the valorization and management of industrial spent catalyst waste in the context of sustainable practice: The case of the State of Kuwait in parallel to European Industry. *Waste Management & Research*, **37**(11), pp. 1127–1141, 2019.

[18] Hogland, W., Marques, M. & Nimmermark, S., Landfill mining and waste characterization: a strategy for remediation of contaminated areas. *Journal of Material Cycle & Waste Management*, **6**, pp. 119–124, 2004.

[19] Jones, P.T. et al., Closing material loops: the enhanced landfill mining concept. *The Journal of the Minerals, Metals & Materials Society*, **64**, pp. 743–744, 2012.

[20] Kamura, K., Omori, M., Kurokawa, M. & Tanaka, H., Estimation of the potential of landfill mining and exploration of metal-enriched zones. *Waste Management*, **93**, pp. 122–129, 2019.

[21] Savage, G.M., Gouleke, C.G. & Stein, E.L., Landfill mining: Past and present. *BioCycle* **34**(5), pp. 58–62, 1993.

[22] Van Passel, S. et al., The economics of enhanced landfill mining: Private and societal performance drivers. *Journal of Cleaner Production*, **55**, pp. 92–102, 2013.

[23] Zhou, C., Fang, W., Xu, W., Cao, A. & Wang, R., Characteristics and the recovery potential of plastic wastes obtained from landfill mining. *Journal of Cleaner Production*, **80**, pp. 80–86, 2014.

[24] Al-Salem, S.M. et al., Potential for thermal and solar power pyrolysis in treating reclaimed real life solid waste from a landfill disposal site. *Proceedings of the 17th International Waste Management and Landfill Symposium (Sardinia'19)*, Sardinia, Italy, 2019.

[25] Vyazovkin, S., Burnham, A.K., Criado, J.M., Pérez-Maqueda, L.A., Popescu, C. & Sbirrazzuoli, N., ICTAC Kinetics Committee recommendations for performing kinetic computations on thermal analysis data. *Thermochim Acta*, **520**, pp. 1–19, 2011.

[26] Vyazovkin, S. et al., *Thermochim Acta*, **590**, pp. 1–23, 2014.

[27] Al-Salem, S.M. & Khan, A.R., On the degradation kinetics of poly(ethylene terephtalate) (PET)/poly(methyl methacrylate) (PMMA) blends in dynamic thermogravimetry. *Polymer Degradation & Stability*, **104**, pp. 28–32, 2014.

[28] Al-Salem, S.M., Al-Nasser, A. & Al-Dhafeeri, A.T., Multi-variable regression analysis for the solid waste generation in the State of Kuwait. *Process Safety & Environmental Protection*, **119**, pp. 172–180, 2018.

[29] Al-Jarallah, R. & Aleisa E., A Baseline study for municipal solid waste characterization for the state of Kuwait. *Waste Management*, **34**, pp. 952–960, 2014.

[30] Mor S., Ravindra K., De Visscher A., Dahiya R.P. & Chandra A., Municipal solid waste characterization and its assessment for potential methane generation: A case study. *Science of the Total Environment*, **371**, pp. 1–10, 2006.

[31] Al-Salem S.M., Leeke G.A., Karam H.J., Al-Enzi R., Al-Dhafeeri A.T., Wang J., Investigating the thermal characteristics of reclaimed solid waste from a landfill site using thermogravimetry. *Proceedings of the 14th International Conference on Environmental Pollution and Pollution Control*, pp. 1273–1276, 2020.

[32] Al-Sarawi, M. A., Massoud, M.S. & Wahba, S.A., Physical properties as indicators of oil penetration in soils contaminated with oil lakes in the Greater Burgan oil fields, Kuwait. *Water, Air & Soil Pollution*, **102**, pp. 1–15, 1998.

[33] Al-Ahmad, M.D.S., Optimization of the assessment and rehabilitation of old landfills in Kuwait. PhD thesis, Rostock University, 2012.

[34] Al-Salem, S. et al., Established Protocols and Research Methods Acquired in The Waste Management Research Unit (WMRU), Technical Report, KISR no. 14854, Project Code: P-KISR-06-11, April, 2018.

[35] Al-Salem, S.M., Gasoline range and chemical wax product recovery from plastic waste reclaimed from a landfill site. *Proceedings of the 11th International Conference on Environmental Science and Development-ICESD*, Barcelona, 2020.

[36] ASTM D 5373, American Society for Testing & Materials, Standard Test Methods for Determination of Carbon, Hydrogen and Nitrogen in Analysis Samples of Coal and Carbon in Analysis Samples of Coal and Coke, Pennsylvania, 2016.

[37] ASTM D 5291, American Society for Testing & Materials, Standard Test Methods for Instrumental Determination of Carbon, Hydrogen, and Nitrogen in Petroleum Products and Lubricants, Pennsylvania, 2016.

[38] ASTM D 5630, American Society for Testing & Materials, Standard Test Method for Ash Content in Plastics, Pennsylvania, 2013.

[39] ISO 1762, Paper, board, pulps and cellulose nanomaterials, Determination of residue (ash content) on ignition at 525°C, 2019.

[40] ISO 11358, Plastics: Thermogravimetry (TG) of polymers, Part 1: General principles, 2014.

[41] Al-Salem S.M. et al., Non-isothermal degradation kinetics of virgin linear low density polyethylene (LLDPE) and biodegradable polymer blends. *Journal of Polymer Research*, **25**, 111, 2018.

[42] Al-Salem S.M. et al., Thermal degradation kinetics of virgin polypropylene (PP) and PP with starch blends exposed to natural weathering. *Industrial & Engineering Chemistry Research*, **56**, pp. 5210–5220, 2017.

[43] Al-Salem, S.M. & Lettieri, P., Kinetic study of high density polyethylene (HDPE) pyrolysis. *Chemical Engineering Research & Design*, **88**(12), pp. 1599–1606, 2010.

[44] Danon B., Mkhize N.M., van der Gryp P. & Görgens, J.F., Combined model-free and model-based devolatilisation kinetics of tyre rubbers. *Thermochimica Acta*, **601**, pp. 45–53.

[45] Dahiya, J.B., Kumar, K., Muller-Hagedorn, M. & Bockhorn, H., Kinetics of isothermal and non-isothermal degradation of cellulose: Model-based and model-free methods. *Polymer International*, **57**, pp. 722–729, 2008.

[46] Holland, B.J. & Hay, J.N., The value and limitations of non-isothermal kinetics in the study of polymer degradation. *Thermochimica Acta*, **388**, pp. 253–273, 2002.

[47] Tuffi, R., D'Abramo, S., Cafiero, L.M., Trinca, E. & Vecchio, S., Thermal behavior and pyrolytic degradation kinetics of polymeric mixtures from waste packaging plastics. *eXPRESS Polymer Letters*, **12**(1), pp. 82–99, 2018.

[48] Blazevska-Gilev J. & Spaseska D., Formal Kinetic Analysis of PVC Thermal degradation. *Journal of the University of Chemical Technology and Metallurgy*, **45**(3), pp. 251–254, 2010.

[49] Taghi Taghizadeh M. & Abdollahi, R., A kinetics study on the thermal degradation of starch/poly (vinyl alcohol) blend. *Chemical and Materials Engineering*, **3**(4), pp. 73–78, 2015.

[50] de Azevedo Gonçalves Mota, R.C., da Silva, E.O. & de Menezes, L.R., Effect of the addiction of metal oxide nanoparticles on the physical, chemical and thermal properties of PVA based nanocomposites. *Materials Sciences and Applications*, **9**, pp. 473–488, 2018.

[51] Kuźnia, M. & Magdziarz, A., Research on thermal decomposition of waste PE/PP. *Chemical and Process Engineering*, **34**(1), pp. 165–174, 2013.

[52] Zong, R., Wang, Z., Liu, N., Hu, Y. & Liao, G., Thermal degradation kinetics of polyethylene and silane-crosslinked polyethylene. *Journal of Applied Polymer Science*, **98**, pp. 1172–1179, 2005.

[53] Agrwal, R.K., Analysis of non-isothermal reaction kinetics. Part 1. Simple reactions. *Thermochimica Acta*, **203**, pp. 93–110, 1992.

DEVELOPMENT OF EFFICACY ASSESSMENT PROCEDURE FOR DISASTER DEBRIS MANAGEMENT IN MUNICIPAL GOVERNMENTS IN JAPAN

IPPEI KAMEDA & NAGAHISA HIRAYAMA
Nagoya University, Japan

ABSTRACT

The objective of this study is to develop an efficacy evaluation method for disaster debris management in municipal governments in Japan. A questionnaire survey was conducted to achieve this goal. The survey used an effort assessment sheet based on Self-Assessment for Business Efforts for Disaster Prevention to assess waste management in municipal governments that are located in the promotion regions of the Nankai Trough catastrophic disaster prevention. The results of this survey indicated that the efficacy in disaster waste management will not only require planning for disaster waste, but also capacity development. Using the Incident Command System, disaster debris management by the municipal governments located in the Nankai Trough earthquake tsunami evacuation measures special enhancement area was found to be more efficient than that by other municipalities. This indicated that the experience of 2011 Tohoku disaster cultivated a sense of ownership about an impending crisis in municipal government officials through the Nankai Trough earthquake disaster prevention activities.
Keywords: disaster debris management, municipal government, efficacy, questionnaire survey, assessment procedure.

1 INTRODUCTION

In Japan, disaster waste management (DWM) is one of the important measures for national land resilience [1]. The Ministry of the Environment (MOE) established the guidelines for the formulation a DWM Plan [2] and implemented model projects [3] in the municipal governments. As a result, municipal governments have been promoting DWM such as formulating DWM Plan. The DWM plan is a plan in which municipal governments organize necessary items for proper, smooth, and prompt disposal of disaster waste before a disaster. And, after the disaster, municipal government will deal with DWM based on the DWM plan [2].

However, municipal governments who had drawn up the DWM Plan before the disaster was confused during the disaster [4], [5]. Therefore, it is necessary to improve the efficacy of municipal governments in DWM not only by planning DWM, but also by continuously check and review the plan already prepared.

Kimura et al. [6] surveyed 12 prefectures that created the DWM Plan after 2011 Tohoku disaster and mentioned the description rates of the items provided in the guidelines for formulating a DWM Plan in the plans prepared by each prefecture. Tajima and Osako [7] are developing a Disaster Resilience Assessment Tool for Waste Management Systems that can be used to improve the ability of municipal officials in carrying out DWM. Municipal governments can use this tool for self-assessing the disaster resilience of its waste management system. This tool uses a rating scale method for assessing items in five stages and aims to manage the progress of DWM by municipal governments. Cutter et al. [8] provided baseline indicators for recovery goals and a methodology for measuring indicators to promote the resilience of local communities. However, so far, little research has been done on methods to assess the efficacy of DWM in municipal governments. Thus, this study aimed to develop an efficacy assessment method for DWM carried out in the municipal

WIT Transactions on Ecology and the Environment, Vol 247, © 2020 WIT Press
www.witpress.com, ISSN 1743-3541 (on-line)
doi:10.2495/WM200181

governments. In addition, this method was also used to assess the efficacy of DWM in designated municipal governments in the Nankai Trough earthquake disaster prevention.

In this study, the efficacy of DWM was defined as, "considering the health of residents, ensuring safety, and taking prompt action for safety and security in terms of hygiene and environment". In addition, improving the efficacy of DWM shall be based on proactive measures. And, the purpose of the Waste Management and Public Cleansing Act in Japan which is a rule before a disaster is required to be adhered to even in DWM.

2 EXAMINATION OF EFFICACY ASSESSMENT METHOD FOR DWM

2.1 Self-assessment item table for business efforts for disaster prevention

The Cabinet Office of Government of Japan published a method for Self-Assessment item table for Business Efforts for Disaster Prevention in 2007 [9]. Using this method, any company of different industry type, business format, and business scale can self-assess its own disaster prevention efforts. This table consisted of 61 items, with major categories as: I. Policy/plan, II. Concrete measures, III. Training, IV. Review, and V. Products and services that contribute to disaster prevention.

As shown in Table 1, the cell value is determined by selecting the implementation level for the target range (width) and positioning (depth) based on the implementation status of the measures of the company. Each company can self-assess their degree of achievement within a fixed level value ranging from 0 to 3 by selecting ABC and abc from "width" and "depth". Furthermore, they can also check the level distribution of achievement from the assessment result of each item and use it to examine the items to be prioritized in the next plan based on the balance of achievement levels of all items.

Table 1: Level value judgment method.

I-b-1. Does your company's management plan include disaster prevention matters?		(Width) Scope of plans including disaster prevention		
		Management plan does not include matters related to disaster prevention.	Management plan includes disaster prevention issues only in certain business areas.	Management plan includes disaster prevention issues in all business areas.
Positioning (depth)		A	B	C
No management plan includes matters related to disaster prevention.	a	0	0	0
Management plan at the department level includes matters related to disaster prevention, but has not been approved by the management.	b	0	1	2
Management plans approved by the management include matters related to disaster prevention.	c	0	2	3

An example of the application of this method in the environmental field is the assessment carried out by Hanshin Water Supply of the enterprise association to examine the future direction of crisis management in water utilities by modifying the items in the table based on the water supply business [10].

2.2 Efficacy assessment method for DWM

In this study, we developed a method for assessing the efficacy of DWM by modifying the Self-Assessment item table mentioned in Section 2.1. It should be noted that the level of efficacy varied depending on the cell value set. Therefore, we considered this aspect and set a reasonable cell value, and we conducted a sensitivity analysis on that value. Furthermore, it was found that higher the level value, higher the efficacy.

2.3 Examination of organizational functioning in DWM

We examined the efficacy assessment of disaster waste countermeasures considering the organizational function at the time of disaster. We considered it effective to calculate the efficacy level values of each function required for DWM to comprehensively improve the DWM by the municipal governments. Tajima et al. [11] systematized the DWM carried out for the Great East Japan Earthquake using the standard crisis response system of the United States of America, Incident Command System (ICS).

Organizational functions required for DWM after a disaster are also important before the disaster. Therefore, in this study, basic functions and sub-functions required for DWM were set using ICS, as shown in Table 2.

Table 2: Basic functions and sub-functions required for DWM.

Basic function	Sub-function	Basic function	Sub-function
Command	Goal setting	Operation	Removal
	Internal adjustment		Storage
	External affairs		Segregation
Public affairs	Public relations		Intermediate treatment
	Human resources		Final disposal
Logistics	Equipment	Planning	Making plan
	Plant		Collecting information
Finance/ administration	Finance		Information analysis
	Contract		Sharing information
	Payment		Tech support
Training		Review	

3 SUMMARY OF QUESTIONNAIRE SURVEY

3.1 Design of the efficacy assessment table

The MOE of Japan mentioned that the management of disaster waste is very important for the conservation of living environment at the time of disaster, and management requires prompt and appropriate initial response based on prior preparations by municipal governments subject of the management [12]. In addition, it has been pointed out that the turmoil immediately after the disaster delays the recovery due to prolonged treatment of disaster waste and increasing treatment costs [13] and significantly impedes subsequent DWM [12]. Therefore, in this study, we focused on the measures to be taken immediately after the disaster.

In this paper, we explain the efficacy assessment table prepared by us based on the self-assessment item table mentioned in Section 2.1.

We first deleted the classifications not applicable to municipal governments from the self-assessment item table, such as "products and services that contribute to disaster prevention" from the major category, and "mutual assistance/mutual assistance with other companies" and "information disclosure/social contribution" from the middle category. Alternatively, "disposal of disaster waste, human waste, etc." was added to the middle category to assess the specific measures for DWM.

The items in "disposal of disaster waste and human waste" were created considering the DWM guidelines and contents of disaster response records issued by the municipal governments and related organizations. As this study focused on measures taken immediately after the disaster, mainly, the "guidance of administrative work in DWM for municipal government" [13], that describes the minimum actions to be taken within 48 hours of the disaster, was referred. Moreover, the deleted items related to information disclosure in "information disclosure/social contribution" were added to this category.

Furthermore, we added "study on cooperation with volunteers" to "Information transmission at the time of disaster, cooperation and coordination with related organizations" and "preparation for business order, business management, etc. for DWM business after disaster" to "Backup and restoration of business facilities and securing financial resources". As a result, the number of items in the efficacy assessment table was 58. We also revised the self-assessment item table to make "width" and "depth" options in each item applicable to DWM. In addition, we interviewed four municipal officials, from different municipal governments with experience in supporting disaster area, on the efficacy assessment table. Table 3 shows the efficacy assessment table.

3.2 Summary of questionnaire survey

A questionnaire survey was carried out with 622 municipal governments in the area of Nankai Trough earthquake disaster prevention, excluding municipal governments in prefectures where a typhoon caused severe damages before the survey. The municipal governments in the region had set specific goals for the planning rate of DWM Plan (60%) and time period for achieving them (by 2025) using the basic plan for Nankai Trough earthquake disaster prevention based on the Act on Special Measures for Promotion of Tohnankai and Nankai Earthquake Disaster Management [14]. The municipal governments were in the process of formulating the DWM Plan. The rate of formulating the plan by the municipal governments in Nankai Trough earthquake disaster prevention was 44% at the end of 2018, higher than the national average of 28% [15]. Therefore, it can be said that the target area in this study was actively engaged in DWM.

In this survey, we mailed the efficacy assessment table to the department in charge of each municipal government on 25 November 2019 and asked them to respond through fax or email by 20 December 2019.

3.3 Questionnaire survey results

Out of 622 municipal governments, 209 responded to the questionnaire and we mention below the status of formulation of DWM Plans by these governments. Survey results

Table 3: Structure of efficacy assessment table and question items.

Category	Subcategory	Question item
I. Policy/plan		
	I-a Policy	
		Municipal comprehensive plan and regional disaster prevention plan include matters related to DWM
	I-b Plan	
		Have a DWM plan
		Measures and implementation of training for safety confirmation are clearly stated in the plan
		Measures and training for Secondary disaster prevention are described in the plan
		Impact of expected disasters is assessed on waste treatment
		At the time of disaster, select priority business for continuity and set its target recovery time
		DWM Plan envisions multiple disaster scenarios
		The plan describes measures and training for emergency recovery in case of facility damage
		Measures and training for initial actions and receiving outside support are described in the plan
		Discuss with related organizations on disaster coordination before disaster
	I-c Organizational structure and command system	
		Personnel with necessary knowledge are secured through experience and training
		Command system at the time of disaster is clearly defined
		Contact means are secured in the event of a disaster
		Command system for nighttime and holidays is in place
II. Concrete measures		
	II-a Ensuring life safety and safety confirmation	
		Procedures for ensuring the safety of executives, staff, and local residents and evacuation methods/routes for emergencies are clear
		Emergency contact network and safety confirmation system are in place
		A response system to prevent secondary disasters is in place
		Storing daily necessities (water, emergency food, and daily necessities) after the disaster
		Implementing measures to confirm the safety of executives and their families
		Implementing measures to return home of employees after a disaster
	II-b Reducing facility damage	
		Implementing earthquake countermeasures (seismic retrofitting) for major facilities
		Implementing storm and flood damage countermeasures for major facilities
		Implementing earthquake countermeasures for facilities and equipment
	II-c Backup and restoration of business facilities and securing financial resources	
		Securing a backup office in case of failure of the government office
		Implementing backup measures for core business systems of major facilities
		Implementing measures for equipment's duplexing (private power generators, etc.)
		Preparing financial resources (subsidies, supplementary budget, insurance, etc.) for disasters
		Preparing for ordering and business management of DWM
		Taking measures (copying etc.) to protect for important documents
	II-d Information transmission at the time of disaster, cooperation and coordination with related organizations	
		Information collection and sharing procedures at the time of disaster are clear
		Means of transmitting information to residents, related organizations, and media at the time of disaster are clear
		Implementing secondary disaster prevention measures to surrounding areas in non-disaster period
		Signing of agreements with local governments and related organizations
		Considering collaboration with volunteers

Table 3: Continued.

Category	Item
II-e Disposal of disaster waste and human waste	
	Actively disclosing information on DWM
	Enlightenment and public information for citizens
	Considering selection and securing of temporary storage site
	Considering management and operation of temporary storage site
	Considering disaster waste processing flow
	Considering hazardous waste and difficult-to-process waste
	Considering the installation of temporary toilets
	Considering household waste, evacuation center waste, and human waste at the time of disaster
	Considering living environment conservation
	Considering progress management of DWM
	Considering environmental measures and monitoring
III. Training	
	Training for emergency contact
	Training on mobilization/gathering
	Training on prevention of secondary disasters
	Training on restoration of facilities (equipment/systems)
	Training on business continuity based on DWM Plan
	Conducting seminars and training for DWM
	Conducting exercises and training for DWM
	Implementation of disaster prevention measures and guidance at home
IV. Review	
	Regular inspection and review of the contents of the plans
	Regular inspection and review of the operation status of the plans
	Regular inspection and review of education and training of DWM
	Regular review of executive officers on their efforts towards DWM
	Third-party diagnosis/audit system

indicated that the planning rate was 72.7% (152 out of 209 municipal governments) and 93.0% (40 out of 43 municipal governments) in the Nankai Trough earthquake tsunami evacuation area located in the Nankai Trough earthquake disaster prevention. Survey results also indicated that many responding municipal governments were formulating DWM Plan and actively working on DWM in the Nankai Trough earthquake tsunami evacuation area.

4 EXAMINATION OF EFFICACY ASSESSMENT METHOD BASED ON QUESTIONNAIRE SURVEY

4.1 Level values in the efficacy assessment table

The cell value in the efficacy assessment table prepared in this study was symmetrically set with respect to diagonal, with the minimum and maximum cell value scores of 0 and 3, respectively, similar to that of the Self-Assessment item table for Business Efforts for Disaster Prevention.

Table 4 shows six combinations of cell values for sensitivity analysis. Using, for example, eqn (1), we calculated the average difference between the raw scores of efficacy level values for each case and case 1 in question k. As a result, the average for 58 solutions was found to be 0.17. Therefore, it can be concluded that the efficacy level value according to the efficacy assessment table was robust. Thus, in this study, we adopted case 1, which was the standard cell value in the matrix of the self-assessment item table

$$\frac{1}{5}\sum_{n=2}^{6}\frac{1}{58}\sum_{k=1}^{58}(\text{raw efficacy value}(n,k) - \text{raw efficacy value}(1,k)). \quad (1)$$

Table 4: Examination case in the sensitivity analysis.

	A × 1	A × 2 (B × 1)	A × 3 (C × 1)	B × 2	B × 3 (C × 2)	C × 3
Case1	0	0	0	1	2	3
Case2	0	0	1	1	2	3
Case3	0	0	1	2	2	3
Case4	0	1	1	1	2	3
Case5	0	1	1	2	2	3
Case6	0	1	2	2	2	3

The average efficacy level value of the 209 municipal governments was calculated and was found to be relatively high in "policy/plan", as shown in Fig. 1. This was because many municipal governments had already formulated the DWM Plan. Alternatively, in "review", the efficacy level values were found to be relatively low. Thus, to ensure the efficacy of DWM Plan, it was considered necessary to inspect and review the plan [13] and maintain and improve the "policy/plan" level values in future. We consider it necessary to improve the efficacy level value of "review". As inspection and reviewing the plan is essential to ensure its efficacy, it was considered necessary to improve the level of "review" in the next. This leads to maintenance and improvement of the level values of "policy/plan".

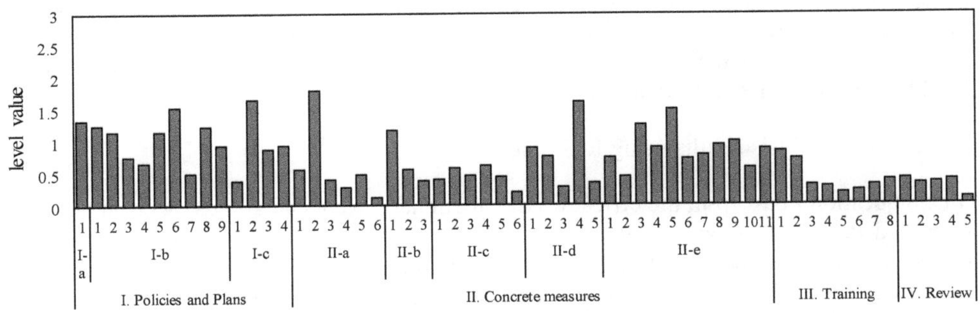

Figure 1: Raw level values of the efficacy.

Alternatively, the level values were found to be low for the items related to facilities and equipment, such as "reduction of facility damage" and "backup and restoration of business facilities and securing financial resources", except for earthquake countermeasures at major facilities. This indicated that promoting hardware measures to execute the DWM Plan is necessary. From human perspective, in "training", the efficacy level values for questions concerning DWM, such as facility restoration related to waste departments and training based on DWM Plan, were lower than those related to disaster prevention, such as mobilization/gathering and emergency contact. Also, in "organizational structure and command system", the level value was low in the item of "personnel with necessary knowledge are secured through experience and training". Therefore, developing human resource for DWM in future is necessary.

In "ensuring life safety and safety confirmation" under "concrete measures", while the efficacy level value of the items in the emergency contact network was high, the level of the items related to evacuation method at the time of disaster, stockpiling of daily necessities, confirmation of safety of the municipal officials' families, and measures taken for those who have difficulty returning home was low. Therefore, studying DWM considering and planning business continuity is necessary.

We next describe the classification of "information transmission at the time of disaster, cooperation, and coordination with related organizations" and "disposal of disaster waste and human waste". In this classification, the level values of the items related to following were high: agreements with related organizations, information gathering, and sharing procedures at the time of disaster; means of transmitting information to residents and disaster waste response; and specific disaster waste measures and human waste measures. First, it was considered that the high level of agreement was because the provision or placement of waste/human waste disposal service was stipulated by the Agreement by the Governor's Association of Japan [16] and that the prefecture had recently signed a comprehensive agreement at the time of disaster with the Industrial Waste Association of prefecture. Next, it was considered that the high level for the items related to DWM and human waste treatment was because 72.7% municipal governments that responded to the survey had already prepared the plan.

Finally, it was considered that the high level for the items related to information collection, information sharing procedures, and means of transmitting information to residents at the time of disaster was because these items were considered during the formulation of the plan. Alternatively, in this classification, the level of questions related to public awareness and public relations during non-disaster period and cooperation with volunteers was low. Therefore, it can be said that voluntary public relations, such as risk communication with residents, cooperation with the social welfare council, and public relations with citizens, should be strengthened during non-disaster period.

4.2 Efficacy analysis results using ICS functions

The 58 question items from the efficacy assessment table were categorized according to ICS functional items, as shown in Table 2, and the average efficacy level value was calculated for each function. While classifying into each function, the content of the question as well as the options were considered. When one question item was applied to multiple sub-functions, it was classified into the two most applicable items. In addition, the question concerning "operation" in the efficacy assessment table referred to multiple sub-functions (removal, storage, sorting, intermediate processing, and final disposal). Thus, sub-functions were not classified, but basic functions were classified as "operation". The main assessment target of this method was municipal governments responsible for the disposal of disaster waste. Therefore, we excluded "tech support" required by countries, prefectures, and experts from the classification. In addition, as there was no question corresponding to the sub-function "payment" in the assessment table, and it was excluded from the classification.

Thus, the efficacy of DWM in the municipal governments was assessed based on eight categories of "command", "public affairs", "logistics", "finance/administration", "operation", "planning", "training", and "review". Fig. 2 shows the assessment results of the efficacy of DWM based on the average level values for municipal governments who responded to the survey in this study.

Figure 2: Assessment of ICS functional efficacy of municipal governments.

In addition, Fig. 2 shows that the efficacy of "operation" was high, while "training", "review", and "finance/administration" were low. Thus, as many municipal governments had already established a DWM Plan, it was considered that they had extensively studied "operation" related to DWM. Alternatively, due to the nature of the plan for disaster countermeasures, verifying the usefulness and validity of the plan for "review" before a disaster was difficult and it was inferred that "review" of the plan was not conducted.

The guidelines of the MOE indicate the importance of checking and reviewing the DWM Plan and mandate municipal governments to predetermine the frequency and method of checking its plans. However, the guidelines do not specifically show the frequency and method of review [2]. Based on the above, we consider that the MOE should provide concrete methods and examples for reviewing the plan. It was speculated that the low level of "training" was because of the following reason: as of 2017, the implementation rate of education and training for DWM in the municipal governments was 10% [17], and the efforts made by municipal governments were not improving. The Fundamental Plan for National Resilience by the Cabinet Office requires the municipal governments to promote the formulation of DWM Plan and develop human resources through education and training to improve their efficacy [17]. In addition, it can be said that many municipal governments did not adequately consider procuring funds and contracts of "general finance". In addition, the Chubu Regional Environment Office of the MOE categorized the items in the DWM Plan of the municipal government in the Chubu block of Japan by dividing them into items before and after the disaster [18]. Kimura et al. [6] investigated the matters to be described in the guidelines prepared by the MOE and the Japan Society of Material Cycles and Waste Management Research, and they arranged them to be described in the DWM Plan. Of these items, those related to "operation" were described, but items related to "finance/ Administration" were not sufficiently described. In other words, it can be said that to improve the efficacy of DWM of municipal governments, it is indispensable to specifically consider matters related to "finance/administration". As a concrete consider matters related to

"Finance/Administration", it was considered that the municipal governments will roughly estimate the DWM cost based on their damage estimation that will be required after a disaster. In addition, after the disaster, affected municipal governments must secure DWM cost until the subsidy is paid by the government. Thus, it is important to inform the finance department in advance of the cost based on damage estimation and adjust the means to secure financial resources. Next, it can be said that it is necessary for the national government to recognize the importance of "finance/administration" and provide specific technical guidance, such as necessary coordination with the finance department and securing financial resources until subsidy issuance to municipal governments.

4.3 Study of the attributes of municipal governments and efficacy assessment

Fig. 3 shows the efficacy assessment results from the perspective of ICS function according to the presence/absence of a DWM Plan in Nankai Trough earthquake tsunami evacuation (Area 1) and Nankai Trough earthquake disaster prevention, excluding Area 1 (Area 2). Fig. 3 shows that the municipal governments that had already prepared a DWM Plan exhibited higher assessment results in all categories than those that had yet not prepared the plan. Therefore, it cannot be said that the efficacy assessment of the DWM does not increase unless a DWM Plan is prepared. However, regarding "training", "review", and "finance/ administration", the efficacy assessment results were low regardless of the status of the formulation of DWM Plan. This was partly because, as indicated earlier, the content of the guidelines to the municipal governments of the MOE and the opportunities for human resource development by the prefectures were not sufficiently created.

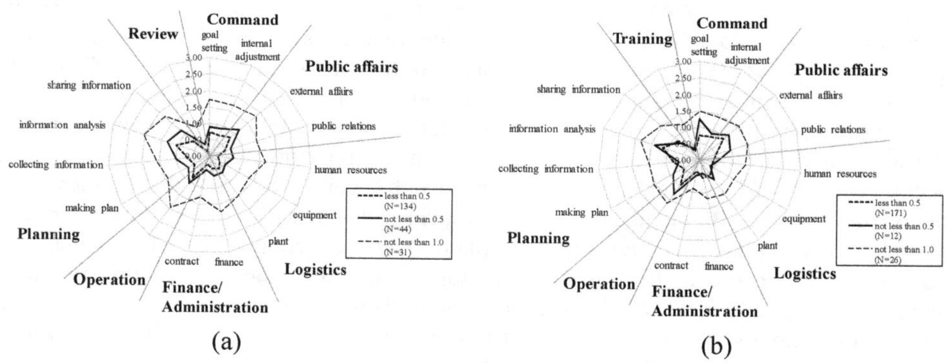

Figure 3: ICS functional efficacy assessment by attribute. (a) Plan; and (b) Area.

Fig. 3 shows that the assessment results of municipal governments of Area 1 were higher in all categories than those of Area 2. It was speculated that this was partly due the formulation of DWM Plan in the municipal governments of Area 1 being higher than that of Area 2. Tajima et al. [19] speculated that fostering a sense of ownership in municipal officials in charge would lead to continuous efforts towards DWM. Furthermore, Karatani et al. [20] found that high school students in the regions of the tsunami disaster prevention who received the tradition of tsunami damage had a higher sense of ownership than those who did not and predicted that tsunami with high waves would come from the port in their towns. This indicated that the experience of 2011 Tohoku disaster would cultivate a sense of ownership in the municipal government officials about the impending crisis through the Nankai Trough earthquake disaster prevention activities.

Fig. 4 shows the efficacy assessment results of the ICS category based on the results of "training" and "review" of the DWM Plan. It can be seen that the efficacy of DWM was high as the assessment results were high, 1.0 or higher in "training", for municipal governments in all categories. In particular, in the items of "human resources", "contracts", and "operation", the efficacy assessment significantly increased. This indicated that "training" is indispensable to ensure the efficacy of DWM and it improves the responsiveness of practical work in the event of a disaster.

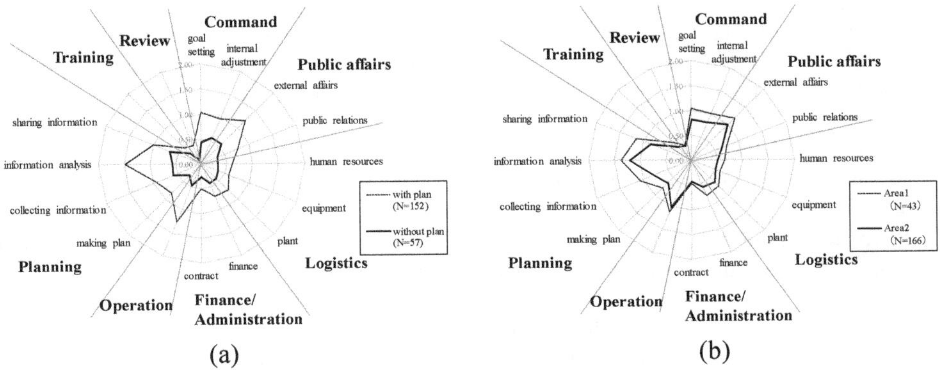

Figure 4: ICS functional efficacy assessment by (a) Training; and (b) Review.

Moreover, it can be seen that as the assessment of "review" increased, the efficacy assessment for all categories improved. However, the efficacy assessment of contracts in "finance/administration" was low, regardless of the assessment result of "review". The bill to revise a part of Waste Management and Public Cleansing Act stipulates certain measures [21], such as simplifying procedures for rapid construction of waste treatment facility and utilizing the waste treatment facilities in the event of a disaster, and alternate processing of disaster waste. However, to improve the efficacy of DWM by municipal governments, it is important to concentrate on activities, such as contracts and procedures at the time of disaster or disaster assessment work.

Thus, we developed and used the method to assess the efficacy of DWM carried out by municipal governments using eight categories: "command", "public affairs", "logistics", "finance/administration", "operation", "planning", "training", and "review". Results indicated that measures and strategies necessary for improving the efficacy of DWM can be clarified at each municipal government. Alternatively, some uncertainties existed in the quality assessment of the described contents because many choices provided for each question of in this efficacy assessment table were provided based on options, such as implementation of measures and presence/absence of description in the plan. Therefore, this table should be customized based on the implementation status of the measures and financial status of the municipal governments who use it for self-assessment. In future, it is necessary to verify the efficacy assessment method before and after disaster response and clarify the relationship between DWM and efficacy assessment.

5 CONCLUSIONS

In this study, we examined the assessment method for the efficacy of disaster waste management (DWM) carried out by the municipal governments by conducting a questionnaire survey in the area of the Nankai Trough earthquake disaster prevention. The findings of the study are described below.

1. Based on the Self-Assessment item table for Business Efforts for Disaster Prevention, an assessment table for the efficacy of DWM in the municipal governments was constructed. We also developed an assessment method using eight categories: "command", "public affairs", "logistics", "finance/administration", "operation", "planning", "training" and "review"
2. It was found that the municipal governments that responded to the survey had extensively studied the "operation" related to DWM. In addition, results also indicated that to ensure the efficacy of DWM, formulation as well as "review" of the DWM Plan and "training" are essential.
3. It was also found that to improve the efficacy of DWM of the municipal governments in Japan, conducting "training" and specifically examining matters related to "finance/ administration" is necessary.
4. It was found that the efficacy assessment method for DWM can clarify the measures and strategies necessary for municipal governments to improve their efficacy.

ACKNOWLEDGEMENTS

I want to thank the municipal officials who cooperated in the questionnaire survey conducted for this study. In addition, I would like to express my gratitude towards the experienced practitioners in disaster response and support who assisted in the design of the survey.

REFERENCES

[1] Cabinet Secretariat Government of Japan, The Fundamental Plan for National Resilience, 2018.

[2] Ministry of the Environment Government of Japan, Disaster Waste Management Guideline (revised edition), 2018.

[3] Ministry of the Environment Government of Japan, Disaster waste management information website. http://kouikishori.env.go.jp/strengthening_measures/municipal_measures/. Accessed on: 7 Jun. 2020.

[4] Ministry of the Environment Government of Japan, How local governments prepare for disposal of disaster waste generated by a giant disaster: What to learn from the case of the Great East Japan Earthquake, 2015.

[5] Kumamoto Prefecture, 2016 Kumamoto earthquake disaster record magazine in Kumamoto city: Toward recovery/reconstruction – 1 year record from the disaster, 2018.

[6] Kimura, K., Nakayama, H. & Shimaoka, T., A study on evaluation by the formulation rate for each item in disaster waste management plan in municipalities. *27th Annual Conference of JSMCWM Lecture manuscript 2016*, pp. 149–150, 2016.

[7] Tajima, R. & Osako, M., Development of disaster resilience assessment tool for waste management system. *28th Annual Conference of JSMCWM Lecture manuscript 2017*, pp. 143–144, 2017.

[8] Cutter, S.L., Burton, C.G. & Emrich, C.T., Disaster resilience indicators for benchmarking baseline conditions. *Journal of Homeland Security and Emergency Management*, 7(1), pp. 1–22, 2010.

[9] Cabinet Office Government of Japan, Self-Assessment item table for "Business Efforts for Disaster Prevention", 2nd ed., 2007.

[10] Hanshin Water Supply, Crisis management measures evaluation and examination support outsourcing report, 2016.

[11] Tajima, R., Hirayama, N. & Osako, M., Identifying emergency support functions for disaster waste management: Structuring the practical tasks undertaken in the case of Great East Japan Earthquake. *J. JSNDS*, **33**, pp. 153–163, 2014.

[12] Ministry of the Environment Government of Japan, Basics of disaster waste management: Learning from past lessons, 2016.

[13] Ministry of the Environment Government of Japan, Guidance of administrative work in disaster waste management for municipal government, 2017.

[14] Cabinet Office Government of Japan, Basic plan for Nankai Trough earthquake disaster prevention, 2019.

[15] Ministry of the Environment Government of Japan, Disaster waste management information website. http://kouikishori.env.go.jp/action/investigative_commission/h30_fiscal_year/pdf/document_h3103_04.pdf. Accessed on:7 Jun. 2020.

[16] The Governor's Association of Japan, Agreement on wide-area support in case of disaster in prefectures nationwide, 2016.

[17] Cabinet Office Government of Japan, *National Resilience Annual Plan 2019*, 2019.

[18] The Chubu Regional Environment Office of Ministry of the Environment Government of Japan, Survey and investigation work report on wide-area disaster waste management in the Chubu block at the time of a large-scale disaster in 2018, 2019.

[19] Tajima, R., Hirayama, N., Mori, T., Kawabata, T., Takata, M. & Osako, M., Effectiveness of workshop training method on development of intentions and attitudes towards disaster waste management preparedness action. *J. JSNDS*, **34**, pp. 99–110, 2015.

[20] Karatani, Y., Koshimura, S. & Shuto, N., Study on systematization of disaster prevention knowledge for sustainable disaster prevention education in tsunami stricken area. *Proceedings of Coastal Engineering*, **50**, pp. 1331–1335, 2003.

[21] Ministry of the Environment Government of Japan, Outline of the bill to revise a part of Waste Management and Public Cleansing Act, 2015.

Author index

www.ingramcontent.com/pod-product-compliance
Lightning Source LLC
LaVergne TN
LVHW062237200926
848484LV00006B/1563
* 9 7 8 1 7 8 4 6 6 4 0 5 3 *